Coral Reefs of the World

Volume 5

Series editors
Bernhard Riegl, Dania Beach, Florida, USA
Richard E. Dodge, Dania, Florida, USA

More information about this series at http://www.springer.com/series/7539

Hajime Kayanne
Editor

Coral Reef Science

Strategy for Ecosystem Symbiosis and Coexistence with Humans under Multiple Stresses

Springer

Editor
Hajime Kayanne
Department of Earth and Planetary Science
Graduate School of Science
The University of Tokyo
Tokyo, Japan

The original version of this book was revised. An erratum to this book can be found at
DOI 10.1007/978-4-431-54364-0_9.

ISSN 2213-719X ISSN 2213-7203 (electronic)
Coral Reefs of the World
ISBN 978-4-431-54363-3 ISBN 978-4-431-54364-0 (eBook)
DOI 10.1007/978-4-431-54364-0

Library of Congress Control Number: 2016943833

Cover image: Bleached coral colony (*Porites cylindrica*) at Kabira Reef, Ishigaki Island in the Ryukyu Islands in 1998. Photo by Saki Harii, University of the Ryukyus.

Printed on acid-free paper

This Springer imprint is published by Springer Nature
The registered company is Springer Japan KK

Preface

Coral reefs comprise a symbiotic system that coexists among coral–algae, ecosystems, landforms, and humans at various levels. They maintain high productions of photosynthesis and calcification: high photosynthetic production forms a basis for the food web, and high calcification constructs coral colonies and coral reef landform to provide habitat for reef creatures. These features explain the highest biodiversity sustained on coral reefs. Humans also benefit from coral reefs with their fishery production and tourism resources and with protection from waves of the outer sea, and in particular with their aesthetic value (Fig. 1).

However, for the last several decades, the symbiotic system of coral reefs has been degraded by local and global human activities. Not only direct destruction such as reclamations and dredges but also silt and nutrient loads from land have degraded coral reef ecosystems. Coral reefs are also degraded by each factor of global warming: ocean acidification by an increase in CO_2, bleaching by warming, and submergence by sea-level rise (Fig. 2).

To pass this valuable ecosystem on to our generations to come, we need to fully understand the symbiotic relationships and their responses to stressors among various levels with biological, chemical, geological, engineering, and historical processes. As these processes are diversified from micro to macro levels, we need to integrate diverse disciplines to achieve this challenge.

The local and global stresses to coral reefs are distinctly represented in Japanese coral reefs as they are mostly the fringing type and are attached to land with increasing development and receive local stressors directly. The coral reefs in Japan are lined in a latitudinal direction and most sensitive to global warming and ocean acidification, which have strong latitudinal gradients.

To integrate different disciplines, we established the Japanese Coral Reef Society in 1997 and hosted the 10th International Coral Reef Symposium in 2004 in Okinawa. We founded a platform for coral reef science; however, approaches were still divided among disciplines. So we launched a program called "Coral Reef Science: Strategy for Ecosystem Symbiosis and Coexistence with Humans under Multiple Stressors" funded by the Ministry of Education, Culture, Sports, Science and Technology, Japan, from 2008 to 2012.

Coral reef science aims to illuminate this complex system and to provide a scientific basis for reconstructing the symbiotic system coexisting between humans and coral reefs. Our multidisciplinary team consisted of 65 researchers in three research areas investigating the present, past, and future of coral reefs. The program consisted of six teams: teams A01 and A02 targeted basic processes of biology and ecology, teams B01 and B02 reconstructed historical change from geological and anthropological points of view, and teams C01 and

C02 took a system approach to evaluate responses to global and local stresses. The disciplines of the teams are diversified, yet we focused on the same questions of responses of coral reefs to local and global stresses and tried to reach a solution.

A description of each team is given below with corresponding chapters in this monograph which review the outcomes of the program. We have understood responses of coral reefs to local and global stresses better than before and have even obtained some tools to predict their responses. Proposals for actual planned countermeasures to sustain healthy reefs are still to come, but we cannot wait long as degradations of coral reefs have been proceeding even during our research activities.

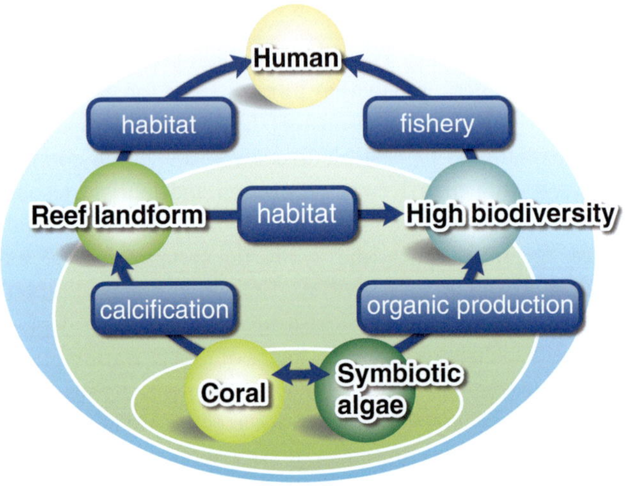

Fig. 1 Multiple scales of symbiotic system among coral-algal symbionts, coral reefs and humans

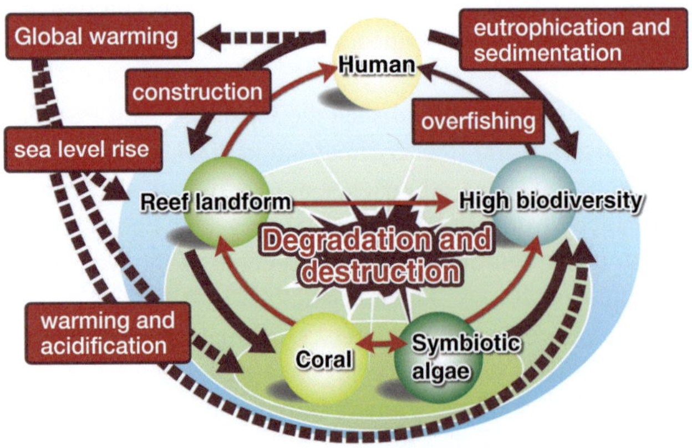

Fig. 2 Degradation of the coral reef symbiotic system by global and local human stressors

A01 Responses of the Coral–Zooxanthella Symbiotic System to Combined Stresses (Chap. 1)

We aimed to understand responses of the coral–zooxanthella symbiotic system to combined stresses. Our major findings were: (1) algal symbionts become a burden for the host under stressful conditions, (2) symbiont type affects the stress response of coral juveniles, and (3) some coral and algal genes exhibit stress-specific changes in expression. Colonial morphology and host genotypes also affect stress susceptibility of corals. The cause of death of corals, changes in recruitment rate, and population dynamics of corals in Okinawa were analyzed.

A02 Linkage Between Biogeochemical Cycles and the Coral Reef Ecosystem (Chap. 2)

The purpose of this research was to understand the dynamic micro ecosystem in coral. We found that 30 °C is a threshold temperature for inducing bleaching and reducing photosynthesis by 50 %. Moreover pathogenic bacteria can enhance and accelerate temperature-mediated bleaching. Primary production in coral reefs is higher than previously thought. New approaches using micro sensors are a "coral holobiont", which is composed of corals with zooxanthellae and associated microbes (bacteria, fungi, and cyanobacteria among others). There is a semi-closed environment for internal coral metabolism.

B01 Historic Changes in Coral Reefs and Their Stressors (Chap. 3)

Based on historic maps, aerial photographs and satellite images at Ishigaki Island, Japan, we discovered that an increase in sediment discharge due to land development in the watershed of the Todoroki River may have caused a decrease in coral cover. Analysis of annual coral bands collected close to the river mouth indicated an increase in sediment discharge and a resulting decrease in coral calcification.

B02 Landscape History of Interactive Relationships Between Human Societies and Coral Reefs (Chaps. 4, 5, and 6)

Viewing a present landscape as a historical artifact of the long-term interactions between two kinds of agents, natural and human, we illustrated holistic landscape histories both of oceanic atolls (so-called low islands) and Ishigaki-jima (a high island) in the Yaeyama Islands in cooperation of geo-archaeology, physical anthropology, historical anthropology, and cultural anthropology, in which various concerns of inhabitants with coral reef were discussed and elucidated in a diachronic perspective from the distant past to the present. We were also aggressive in delivering outreach activities making the most of our academic outcomes in order to meet the challenge of shared interests among a variety of the present inhabitants in their future coexistence with coral reefs.

C01 Responses of Coral Reefs to Global Warming (Chap. 7)

Responses of coral reefs to warming, ocean acidification, and sea-level rise were investigated and evaluated with community and ecosystem scales in the field. Bleaching induced by high temperatures occurred every several years, initiating degradation of coral reefs. Effects of bleaching and acidification varied by species, and only a few species responded to sea-level rise. A shift from hermatypic coral to macro-algae had been assumed, but an example of a shift to soft coral by ocean acidification was seen.

C02 Comprehensive Assessment and Prediction of Multiple Stresses and Model Analysis of Reef Ecosystem Response (Chap. 8)

With the various associated field surveys, a comprehensive model system was developed to evaluate the propagation processes of multiple environmental stresses on a reef ecosystem and to quantitatively describe the carbonate system dynamics and organic matter and nutrient dynamics in the reef. Additionally, as one of the most critical and novel models for the reef ecosystem response modeling under multiple environmental stresses, a "coral polyp model" was successfully developed. The processes to yield the terrestrial environmental loads were analyzed in relation to the socioeconomic aspects of the local society, which is crucial for establishing an effective management scheme of man-made environmental stresses.

Tokyo, Japan Hajime Kayanne

Contents

Life History and Stress Response of Scleractinian Corals

1

Michio Hidaka

Abstract

Symbiosis with zooxanthellae (symbiotic dinoflagellates) and high capacities for clonal reproduction and regeneration are important life history traits of reef-building corals. This chapter reviews the life history of reef-building corals and the symbiotic relationships between corals and their algal symbionts and discusses possible collaborative defense systems against environmental stresses as well as mechanisms of adaptation to environmental changes in coral–zooxanthella symbiotic systems. Most corals associate with one main type of symbiont, although minor or cryptic symbiont types are often detected within a colony. The association between the coral host and algal symbiont appears to be stable, and the original symbiont often returns when corals recover from bleaching. Some corals, such as the massive *Porites*, exhibit high fidelity to certain types of algal symbionts and still have high stress tolerance as well as long life spans suggesting the possibility that stem cells and algal symbiont cells in such long-lived colonies accumulate mutations; in turn, if cells with a higher fitness for a new environment proliferate within a colony, the colony may be able to adapt to the new environment.

Keywords

Life history • Planula • Symbiosis • Stress response • Zooxanthella

1.1 Life History and Symbiosis in Corals

"Coral" describes a group of cnidarians that secrete a calcium carbonate skeleton. Hermatypic corals (reef-building corals) are those associated with symbiotic dinoflagellates, which are generally called zooxanthellae. Hermatypic corals generally secrete skeletons faster than ahermatypic corals, which are not associated with symbiotic dinoflagellates. Thus, hermatypic corals are a major contributor to the formation of coral reefs.

1.1.1 Life Cycle of Cnidarians: Polyp and Medusa

Cnidarians possess two body plans, the polyp and the medusa. The polyp is a sessile stage attached to a substrate on the seafloor, while the medusa is a free-swimming stage. Despite the marked difference in appearance, polyps and medusas are topologically the same, and both body plans are composed of a bag with one opening called a mouth. The wall of the bag consists of two cell layers, the epidermis and the gastrodermis, separated by mesoglea, which is composed of extracellular materials such as collagen. Because cnidarians have two cell layers derived from ectoderm and endoderm, they are considered animals with two germ layers and hence basal metazoans.

The phylum Cnidaria consists of four classes, Anthozoa, Hydrozoa, Cubozoa, and Scyphozoa (Fig. 1.1). Hydrozoans,

The original version of this chapter was revised. An erratum to this chapter can be found at DOI 10.1007/978-4-431-54364-0_9.

M. Hidaka (✉)
University of the Ryukyus, Nishihara, Okinawa 903-0213, Japan
e-mail: hidaka@sci.u-ryukyu.ac.jp

H. Kayanne (ed.), *Coral Reef Science*, Coral Reefs of the World 5, DOI 10.1007/978-4-431-54364-0_1

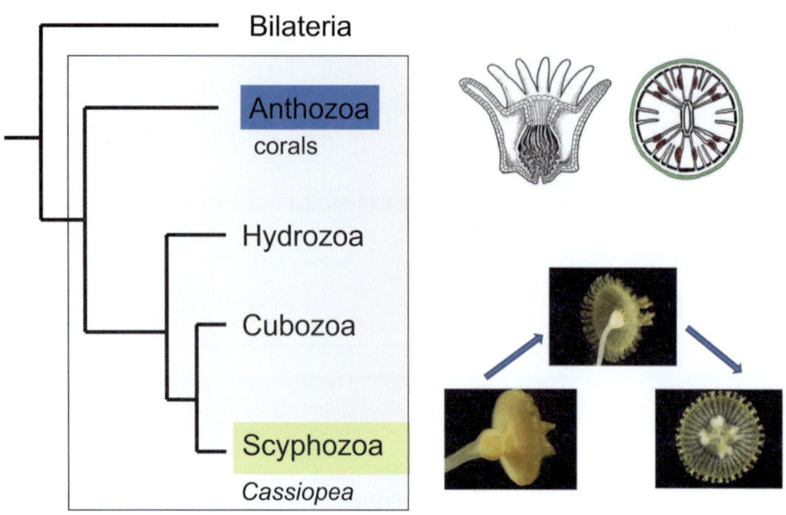

Fig. 1.1 Phylogenetic relationship of four classes of Cnidaria (After Miller et al. 2005). *Upper right*: body pattern and cross section of a coral polyp (Illustration by Dwi Haryanti). *Lower right*: medusa formation via monodisk strobilation in *Cassiopea* jellyfish

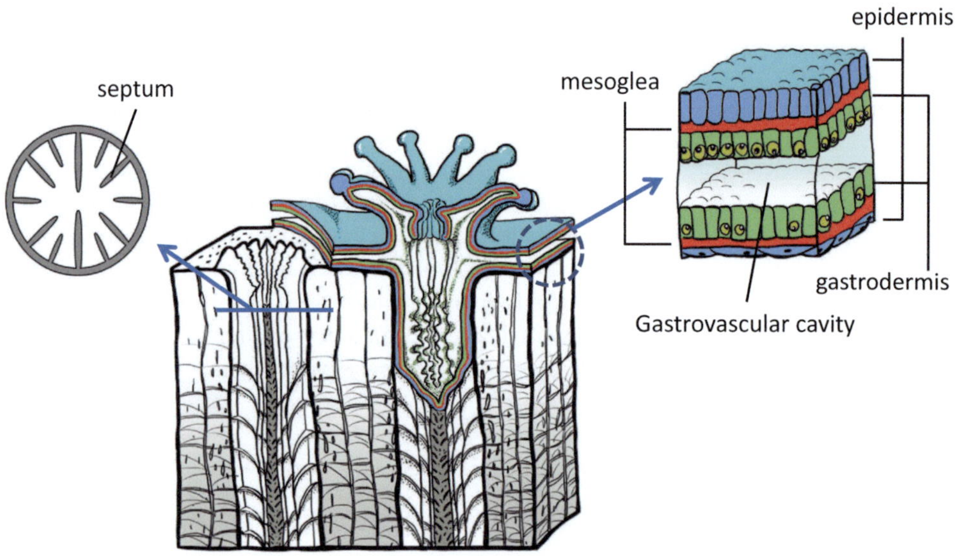

Fig. 1.2 Diagrammatic scheme of tissue and skeletal structures of a part of a coral colony (Illustration by Dwi Haryanti)

cubozoans, and scyphozoans generally pass through both polyp and medusa stages in their life cycle; thus, these three classes are called Medusozoa. In Medusozoa, polyps form medusas asexually, and medusas produce gametes and undergo sexual reproduction, although exceptions exist (e.g., hydra). Sexually produced larvae settle to the substratum and metamorphose into polyps. In Anthozoa, the medusa stage is absent, and sexual reproduction is performed by polyps. The pattern of medusa formation differs among the three classes of Medusozoa. In Hydrozoa, the medusa is formed via budding from the lateral wall of polyps, while in Cubozoa, an entire polyp metamorphoses into a medusa. In Scyphozoa, the medusa is formed by transverse fission of polyps in a process called strobilation.

1.1.2 Body Structure of Corals

The polyp structure of scleractinian corals is similar to that of sea anemones, a bag with a mouth or oral opening surrounded by a ring of tentacles. Most corals are, however, colonial, and polyps within a colony are connected by a common tissue layer called the coenosarc. The gastrovascular cavity of each polyp within a colony is also connected by a common gastrovascular canal system within the coenosarc. The aboral body wall secretes a skeletal structure such as the corallite, a cuplike structure that houses a polyp, and the coenosteum, which underlies the coenosarc. The coral colony can be considered a bag with multiple mouth openings covering the skeleton (Fig. 1.2).

1.1.3 Life History Traits of Corals

This section reviews recent research on coral life history, especially in terms of sexual and asexual reproduction, the acquisition of symbiotic dinoflagellates, and aging and life spans of corals.

1.1.3.1 Sex of Corals and Formation of Reproductive Cells

Approximately 70–75 % of corals are hermaphroditic, while the remaining 25–30 % are gonochoric (Harrison 2011; Kerr et al. 2011). In hermaphroditic corals, oocytes and testes develop within the same polyp, while in gonochoric corals, male colonies (individuals in the case of solitary corals) produce sperm and female colonies (or individuals) produce oocytes. About 87–90 % of corals release sperm and/or eggs to seawater, and fertilization occurs within the seawater column (Harrison 2011; Kerr et al. 2011). These corals are called broadcast spawners (Fig. 1.3). The planula larvae of broadcast-spawning corals develop within the first 24 h and generally settle onto substrate within 3–5 days of spawning (Babcock and Heyward 1986). In the other 10–13 % of corals, fertilization occurs within the polyp, and fertilized eggs develop to planula larvae in the gastrovascular cavity of the polyp; these are called brooders (Fig. 1.3). Observing fertilization within the polyp in brooder corals is difficult, but genetic evidence suggests that offspring acquire paternal genes in some brooder corals (Ayre and Resing 1986; Yeoh and Dai 2010). Early developmental stages are observed in some brooding species such as *Favia fragum* (Szmant-Froelich et al. 1985), *Isopora palifera* and *Isopora cuneata* (Kojis 1986), and *Pocillopora damicornis* (Permata et al. 2000).

A small number of species are reported to display both modes of reproduction, brooding and spawning. Harrison (2011) stated that strong systematic trends can be observed in sexual patterns, the arrangement of gametes in polyps, and sperm structure, whereas the mode of reproduction is more variable and may be a plastic life history trait. Kerr et al. (2011) reconstructed evolutionary transitions in sexual patterns and reproductive modes based on the modern molecular phylogeny of scleractinian corals. They demonstrated that the ancestor of scleractinian corals was gonochoric and that hermaphrodites have independently arisen in three distantly related lineages. Kerr et al. (2011) also suggested that the most probable pathways of state changes are from gonochoric spawner to gonochoric brooder, then to hermaphroditic brooder, and finally to hermaphroditic spawner.

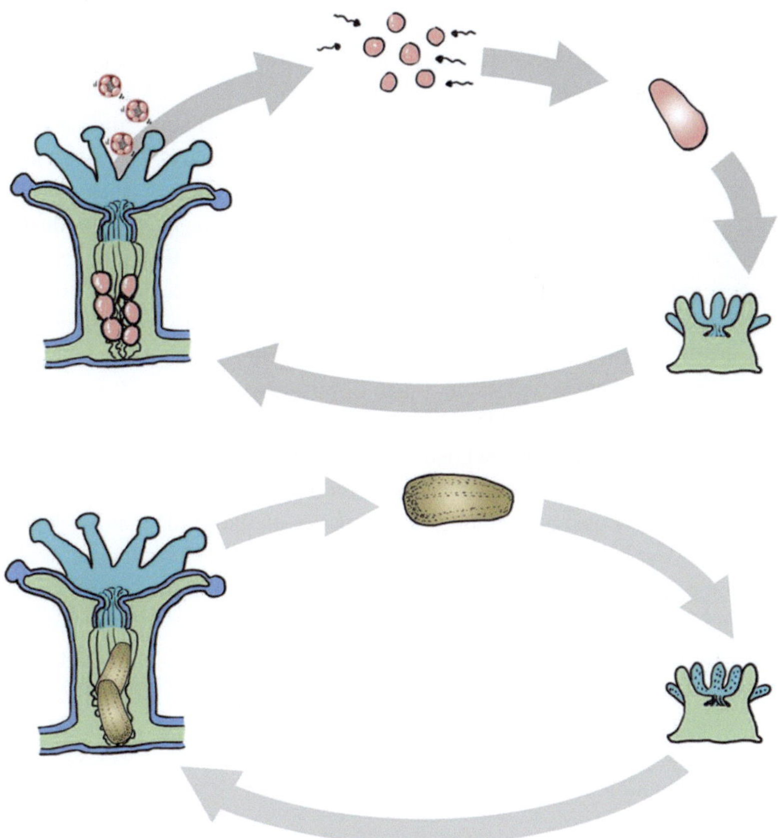

Fig. 1.3 Reproductive modes of scleractinian corals. *Upper*: spawning coral releasing egg–sperm bundles. *Lower*: brooding coral releasing planula larva (Illustration by Dwi Haryanti)

In both spawners and brooders, planula larvae metamorphose to polyps after settlement. The first polyp is called the primary polyp. In colonial corals, the extrathecal tissue of the primary polyp grows outward, and new polyps are formed by budding or via intratentacular budding.

Gonads develop within mesenteries or in the region protruding from mesenteries. Germ cells of sea anemones originate in the gastrodermis in the mesentery but move into the mesenterial mesoglea where they mature (reviewed by Fautin and Mariscal 1991). Recently Shikina et al. (2012) showed that the earliest stage of oocytes of the coral *Euphyllia ancora* was located along the mesenterial mesoglea, while oocytes of more developed stages occur in the mesenterial mesoglea. Sex determination and control of germ cell differentiation in corals require further study. One possible avenue of future research involves DM domain-containing protein, which is important for sex determination in other metazoans such as flies, nematodes, and humans. Miller et al. (2003) identified a gene, *AmDM1*, from the coral, *Acropora millepora*, that encodes a homologous DM domain-containing protein. This species differentiates sexual cells seasonally, and *AmDM1* transcripts are present at higher levels in branch tips collected in October than in samples collected during other months. Miller et al. (2003) suggested that *AmDM1* might be involved in sexual differentiation, which may occur seasonally before the spring spawning. Sex chromosomes have been suggested to exist in *Acropora solitaryensis* based on comparative genomic hybridization using DNA of sperm and unfertilized eggs (Taguchi et al. 2014).

Upregulation of *cryptochrome 2* (*cry2*) may play a role in the timing of spawning of *A. millepora*, as *cry2* expression is increased by full moonlight (Levy et al. 2007). How upregulation of *cry2* leads to the initiation of spawning in this coral remains to be determined. In a brooding coral, *F. fragum*, however, the pattern of *cry1*, *cry2* expression did not correlate with monthly planula release events, suggesting that these genes are not involved in the entrainment of reproductive cycles to lunar light cycles in this species (Hoadley et al. 2011).

Twan et al. (2006) reported that the concentrations of sex hormones, such as testosterone, estradiol, gonadotropin, and gonadotropin-releasing hormone, in the tissue of the coral *Euphyllia ancora*, increase during the breeding season. Furthermore, the activity of aromatase, which is involved in synthesizing estradiol from testosterone, also increases during the breeding season of this coral (Twan et al. 2006). Thus, sex hormone is likely responsible for the maturation of germ cells and initiation of spawning. How the maturation of germ cells is regulated by environmental factors needs further examination.

1.1.3.2 Acquisition of Symbionts by Coral Juveniles

Most spawning corals produce gametes without zooxanthellae. Offspring of these corals must acquire zooxanthellae from the environment at the planula or primary polyp stage. This mode of symbiont acquisition is called horizontal transmission. Spawning species of the genera *Montipora*, *Porites*, and *Pocillopora* release eggs containing zooxanthellae (Fig. 1.4). In these species, zooxanthellae enter the oocytes before they are spawned. Most brooder corals release zooxanthellate planulae. Corals that inherit their symbionts from their mother are called vertical transmitters. Only a small number of brooding corals release zooxanthella-free larvae (e.g., *I. palifera* and *I. cuneata*; Kojis 1986).

1.1.3.3 Pattern of Zooxanthella Localization to the Gastrodermis During Development

In corals with the vertical mode of symbiont transmission, zooxanthellae become restricted to the gastrodermis as fertilized eggs develop to planula larvae. The timing and mechanism of zooxanthella localization differ among *Porites*, *Montipora*, and *Pocillopora* (Hirose and Hidaka 2006; Fig. 1.5). Zooxanthellae are restricted to the animal hemisphere of the oocyte in *Pocillopora verrucosa* and *Pocillopora eydouxi*, whereas they are distributed evenly in the oocytes of *Porites cylindrica* and *Montipora digitata* (Fig. 1.5). In *P. verrucosa* and *P. eydouxi*, zooxanthellae are delivered to some blastomeres, whereas they are delivered to all blastomeres during cleavage in *P. cylindrica* and *M. digitata*. In *P. cylindrica*, *P. verrucosa*, and *P. eydouxi*, blastomeres containing zooxanthellae drop into the blastocoel and differentiate into gastrodermal cells during gastrulation, which leads to zooxanthella localization in the gastrodermis at the gastrula stage. Marlow and Martindale (2007) suggested that parts of blastomeres containing zooxanthellae drop into the blastocoel and are phagocytosed by gastrodermal cells in *Pocillopora meandrina*. In contrast, zooxanthellae are distributed in both gastrodermis and epidermis in the gastrula of *M. digitata*. Zooxanthellae remaining in the epidermis might move to the gastrodermis through the mesoglea as the planulae mature. Similarly, in a brooder coral *Euphyllia glabrescens*, zooxanthellae are distributed mainly in the epidermis of planulae when the planulae are released, but, later, zooxanthellae become restricted to the gastrodermis as the planulae develop (Huang et al. 2008). Whether an entire zooxanthella-containing cell migrates to the gastrodermis or if only the zooxanthellae are translocated from the epidermis to the gastrodermis remains unclear.

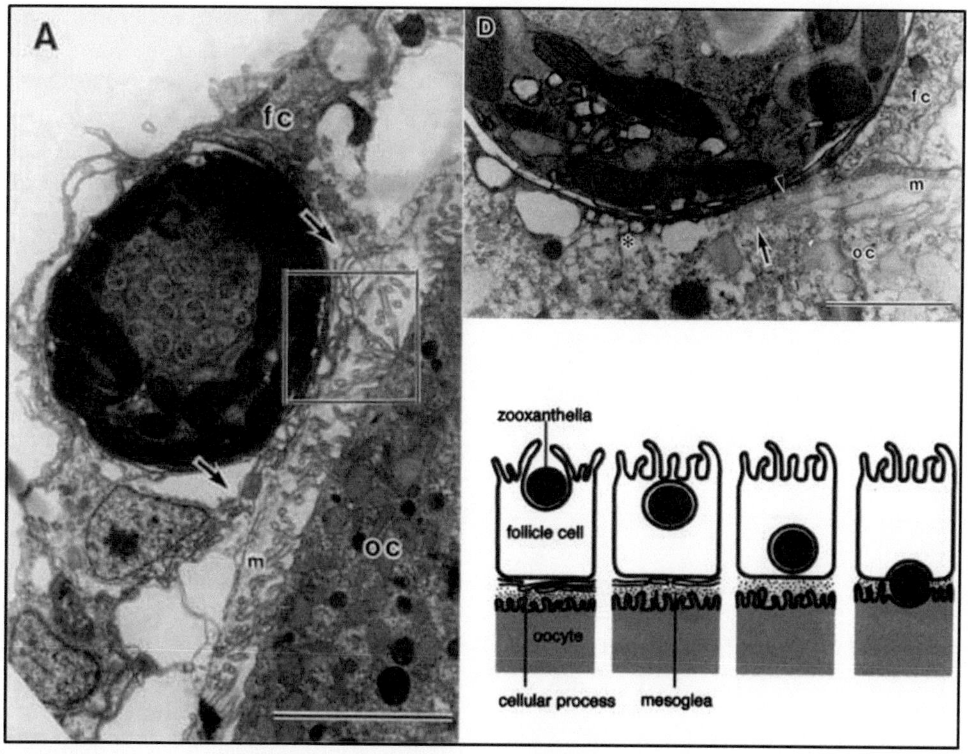

Fig. 1.4 Entry of *Symbiodinium* cell into oocyte of *Pocillopora verrucosa*. Zooxanthellae are translocated from the follicle cell to oocyte through a temporarily opened gap in the mesoglea (With kind permission from Springer Science + Business Media: Coral Reefs 20: 273–280, Hirose et al. (2001) Figs. 4 and 6)

In the horizontal transmitter corals *Acropora* spp., larvae were historically thought to be unable to take up zooxanthellae and that zooxanthella acquisition occurred at the primary polyp stage (e.g., Hirose et al. 2008a). However, Harii et al. (2009) demonstrated that the planulae of several species, including *Acropora* spp., can acquire zooxanthellae when they are artificially inoculated with homologous (zooxanthellae isolated from the same species) or heterologous (zooxanthellae isolated from different species) zooxanthellae. Planula larvae begin to take up zooxanthellae when they develop an oral opening and gastrovascular cavity, i.e., 5 and 6 days after fertilization in *Acropora tenuis* and *Acropora digitifera*, respectively (Figs. 1.6 and 1.7). The acquisition of zooxanthellae and onset of symbiosis in larvae of *Fungia scutaria* have also been reported (Schwarz et al. 1999).

In the field, zooxanthellae are present in the seawater column as well as in sediments and are thus available for acquisition by coral larvae (Yamashita and Koike 2013). Larvae of *Acropora monticulosa* take up zooxanthellae if sediment is added to a chamber containing filtered seawater and larvae (Adams et al. 2009). Planula larvae of *A. tenuis* and *Favia pallida* can take up zooxanthellae in the field when kept in bottles with a mesh window (Harii et al. unpublished observation). These observations strongly suggest that coral larvae can acquire zooxanthellae in the field. However, how efficiently planulae take up zooxanthellae during the planktonic phase remains to be investigated.

Possessing symbiotic dinoflagellates may be beneficial for planulae, as the symbiotic algae could provide host planulae with photosynthetic products (Harii et al. 2010). However, symbiotic algae can also be a source of reactive oxygen species (ROS) when exposed to high temperature and strong light stress at the ocean surface (Yakovleva et al. 2009; Nesa et al. 2012, see Sect. 1.3.1). If zooxanthella acquisition occurs after settlement at the primary polyp stage, larvae avoid oxidative stress during the planktonic phase and may acquire zooxanthellae adapted to the environment where they settle. This may be one reason why most corals are horizontal transmitters, whose offspring acquire symbiotic algae from the environment at each generation.

1.1.3.4 Asexual Reproduction, Senescence, and Life Span (Longevity)

Most colonial corals have a high capacity of regeneration and form new colonies from fragments when a colony is broken into pieces by physical forces such as wave action. This mode of asexual reproduction in corals is called fragmentation (Highsmith 1982; Smith and Hughes 1999). Because their branches are easily broken by wave action,

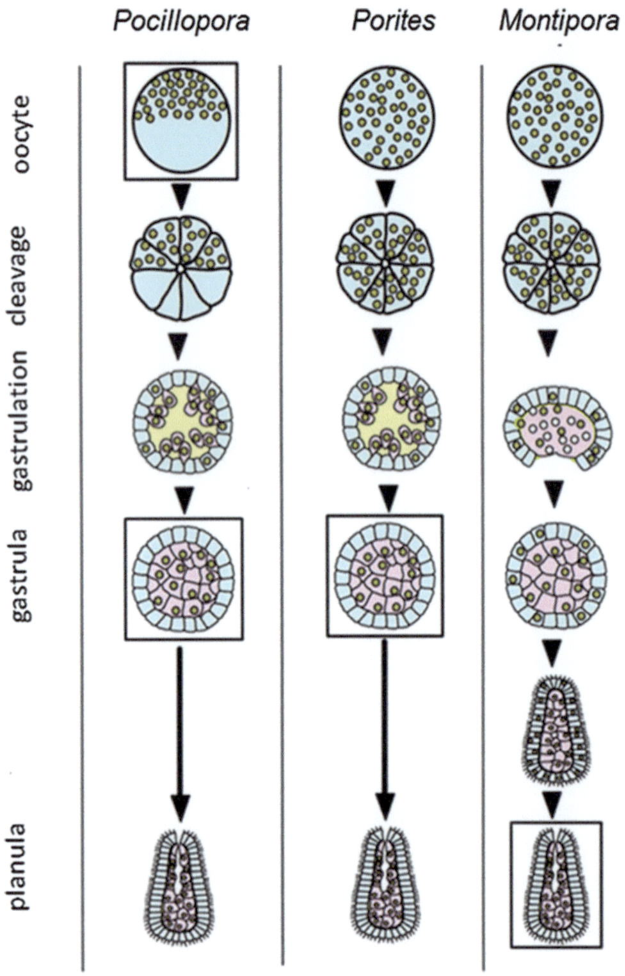

Fig. 1.5 Process of localization of algal symbionts to the gastrodermis during early development of three corals with vertical mode of symbiont transmission. Changes in the distribution pattern during development are compared among *Pocillopora verrucosa*, *Porites cylindrica*, and *Montipora digitata* (From Fig. 4 of Hirose and Hidaka (2006), *Zoological Science*, Vol. 23, No. 10: 873–881)

branching corals tend to reproduce asexually via fragmentation. This is also true of massive *Galaxea fascicularis* colonies, some morphs of which have a mechanically weak, vesicular coenosteum (Wewengkang et al. 2007). Solitary corals belonging to the genera *Fungia* and *Diaseris* undergo asexual reproduction by dissolving certain parts of their skeleton (Yamashiro et al. 1989). In *Fungia* corals, disk detachment is facilitated by skeletal dissolution along a certain plane of the stalk, and even weak mechanical force can separate the disk from the stalk at a predetermined plane. Individuals of *Diaseris fragilis* are circular in shape, but they break their body into several sectors, and each sector regenerates to form circular individuals. Under stress, polyps dedifferentiate into planula-like polyp balls that settle and metamorphose into polyps. This phenomenon, called polyp bailout, has been reported for *Seriatopora hystrix* and several other branching corals (Sammarco 1982). Subsequently, the metamorphosed polyps may form regenerated colonies. Even small piece of coral tissue can regenerate into polyps (Kramarsky-Winter and Loya 1996; Vizel et al. 2011). *P. damicornis* planulae are believed to be produced asexually (Stoddart 1983), although at least some planulae develop from fertilized eggs, and early developmental stages of *P. damicornis* planulae have been observed in an Okinawan population (Permata et al. 2000). A recent study using microsatellite markers demonstrated that *P. damicornis* produces both asexual and sexual planulae (Yeoh and Dai 2010).

Colonies that are asexually produced from a source colony are genetically identical to the source colony; hence, they are considered to be clonemates with one another. Coral populations generally consist of clonal colonies derived from asexual reproduction and non-clonal colonies with unique genetic composition derived from sexual reproduction. The relative contribution of sexual and asexual reproduction to the maintenance of coral populations has attracted the interest of many researchers.

1.1.3.5 Coloniality and Life Span of Corals

The life span of asexual lineages of a coral colony is currently unknown. In fact, the longevity of corals in general is not well understood. Massive *Porites* colonies may live for more than hundreds of years (Hughes and Jackson 1985; Potts et al. 1985; Lough and Barnes 1997), and the ages of these colonies are estimated based on the size of the colony and their annual growth rate or annual growth bands. Estimating the age of branching corals is more challenging, as they may reproduce asexually via fragmentation. Whether colonies regenerated from fragments are of the same age as their source colony is unknown.

Over a 1-year observation period, Permata and Hidaka (2005) demonstrated that colonies regenerated from adult branch tips of *P. damicornis* did not form branches but instead grew horizontally. The regenerated colonies appeared to either lose the capacity or exhibit reduced capacity to originate branches due to aging (Permata and Hidaka 2005), suggesting that colonies regenerated from fragments may be of the same age as their source colonies. In contrast, polyps regenerated from small pieces of tissue of the solitary coral *Fungia* sp. appear to be rejuvenated. If two regenerated polyps were brought into contact, they fused with each other (Fig. 1.8), whereas two individuals of adult *Fungia* never fused when their intact surfaces touch each other (Jokiel and Bigger 1994). Occasionally, more than one polyp regenerated from a piece of tissue to form a colonial *Fungia* (Fig. 1.8). Thus, newly regenerated polyps of *Fungia* appeared to display colonial characteristics. However, when they formed a disk, the disk edges of two individuals did not fuse (Fig. 1.8), suggesting that at a certain stage of development (or regeneration), *Fungia* corals acquire

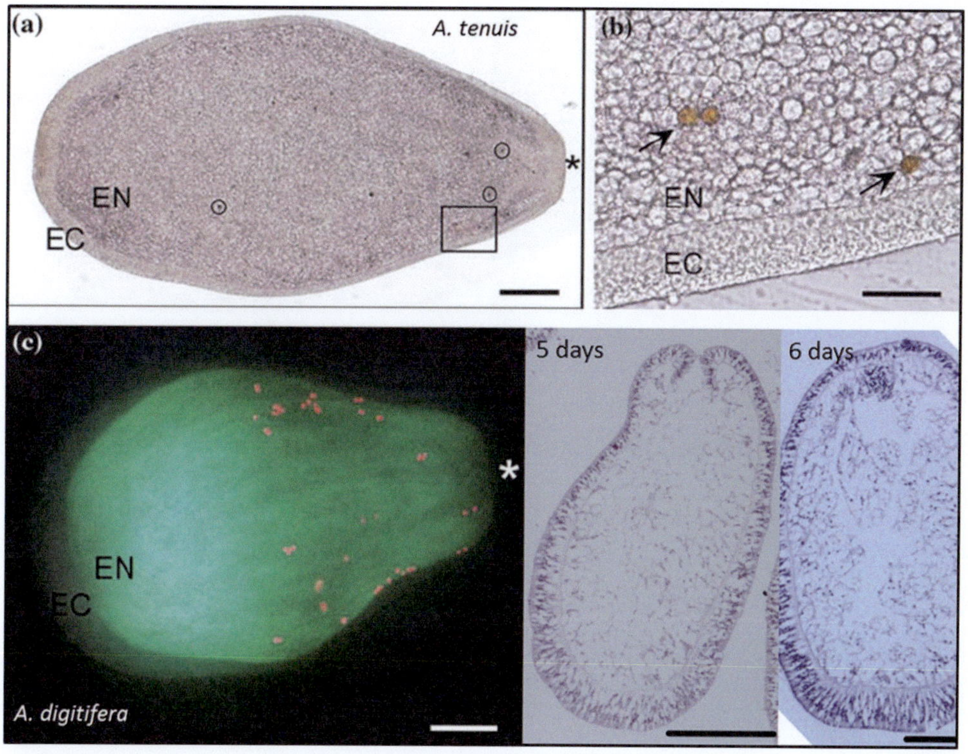

Fig. 1.6 Planulae of *Acropora tenuis* and *A. digitifera* inoculated with homologous symbionts, that is, freshly isolated zooxanthellae from colonies of the same species. Planulae take up zooxanthellae when mouth opening and gastrovascular cavity are formed 5 or 6 days after fertilization (With kind permission from Springer Science + Business Media: Marine Biology 156: 1203–1212, Harii et al. (2009) Figs. 1 and 3)

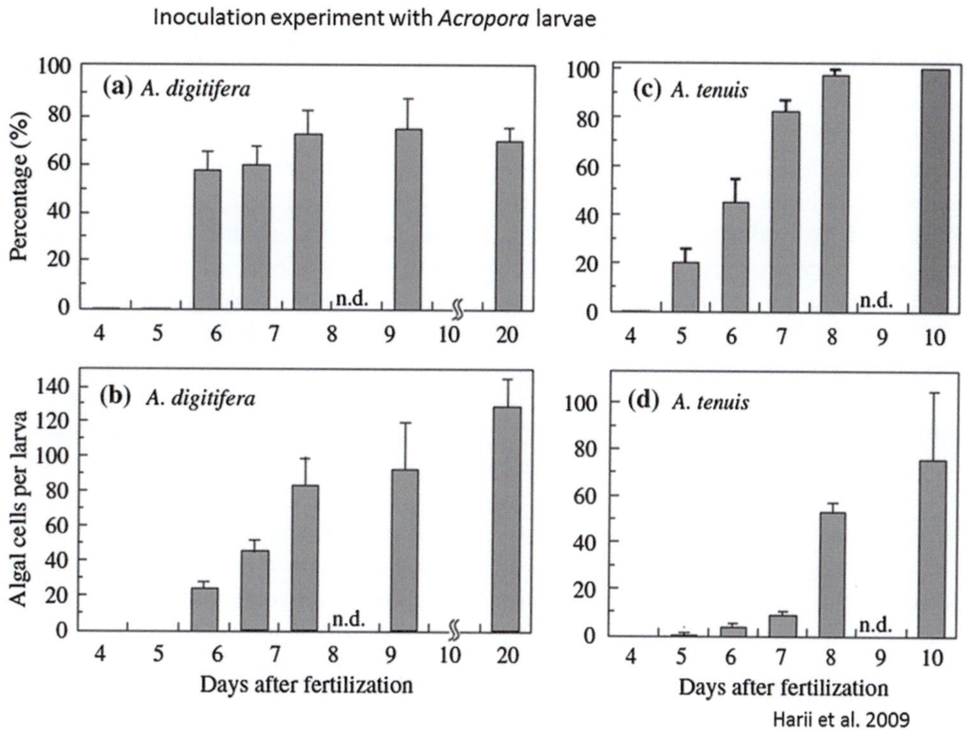

Fig. 1.7 Success ratio of inoculation experiment with *Acropora tenuis* and *A. digitifera* planulae. The percentage of planulae that took up zooxanthellae and the number of algal cells per infected larva on each day. Planulae 4 or 5 days after fertilization did not take up zooxanthellae but started to take up zooxanthellae 5 or 6 days after fertilization, in *Acropora tenuis* and *A. digitifera*, respectively (With kind permission from Springer Science + Business Media: Marine Biology 156: 1203–1212, Harii et al. (2009) Fig. 2)

Fig. 1.8 (**a**) Three polyps regenerated from a small piece of tissue of *Fungia* sp. 162 days after isolation of the tissue piece. (**b–d**) Contact experiment with polyps regenerated from tissue pieces derived from the same individual of *Fungia* sp. (**b**) Commencement of contact experiment. (**c**, **d**) Fusion of the polyps shown in (**b**) 100 days after contact. *Side* and *top* views

solitary characteristics. This observation points to the possibility that some corals undergo rejuvenation when regenerated from a small piece of tissue. Whether coloniality is the ancestral character in scleractinian corals remains to be determined. Most basal metazoan phyla have representatives that reproduce asexually to form clones (Nilsson Sköld and Obst 2011), indicating an ancestral origin of agametic cloning. However, studies on the evolutionary distribution of agametic cloning have suggested that coloniality has evolved several times within different phylogenetic lineages of invertebrates (Geller et al. 2005; Nilsson Sköld and Obst 2011).

1.1.3.6 Estimation of Coral Age

If a nondestructive method could be developed to estimate the age of corals using soft tissue, the technique would provide a powerful tool to study the demography, senescence, and possible rejuvenation of corals during asexual reproduction. Furthermore, a method to estimate coral age based on measurements of telomere length would also offer a useful approach to study population dynamics and life history traits such as aging and longevity of corals.

Telomeres are tandem repeat sequences of DNA at the ends of eukaryotic chromosomes. Generally, telomeres shorten during each cell division due to incomplete replication of the end of double DNA strands; thus, telomere length may reflect the age of some animals. Single telomere length analysis (STELA), which is used to estimate the length of single telomeres of specific chromosomes, has been applied to corals. Ojimi et al. (2012) found that sperm of the solitary coral *Ctenactis echinata* exhibit longer telomere lengths compared to somatic tissues. In contrast, Tsuta and Hidaka (2013) found no significant difference in telomere length among sperm, planula larvae, and polyps of adult colonies using STELA. Although conventional terminal restriction fragment (TRF) analysis has shown that sperm telomeres tend to be longer than those of polyps, the difference was not significant. However, they did observe significant differences in the mean TRF length among sperm, larvae, and adult colonies in *A. digitifera*; in this species, the mean TRF length shortened during development (Tsuta et al. 2014).

To better understand the cellular mechanisms underlying the high capacity for regeneration and longevity of corals (as is known in hydra), one must first examine the distribution of somatic stem cells or stemlike cells and their maintenance and differentiation into somatic cells (Watanabe et al. 2009; Gold and Jacobs 2013). Such information regarding the high regeneration capacity and its cellular mechanisms would contribute to the establishment of efficient reef restoration.

1.2 Symbiosis Between Corals and Zooxanthellae

1.2.1 Diversity of Zooxanthellae

1.2.1.1 Classification of Zooxanthellae

Until the mid-1980s, zooxanthellae associated with diverse marine invertebrates such as foraminifera, coral, and bivalves like *Tridacna* were all considered to belong to the same species, i.e., *Symbiodinium microadriaticum*. Trench and Blank (1987) determined the number of chromosomes of zooxanthellae based on a three-dimensional reconstruction of transmission electron microscopy (TEM) images of serial thin sections of their nuclei and documented that zooxanthellae isolated from four different host species harbored different numbers of chromosomes. The numbers of chromosomes within the zooxanthellae were 26, 50 $\pm$ 1, 74 $\pm$ 2, and 97 $\pm$ 2 [means $\pm$ SD] in symbionts derived from the coral *Montipora verrucosa*, the sea anemone *Anthopleura elegantissima*, the sea anemone *Heteractis lucida*, and two species of the jellyfish *Cassiopea xamachana* and *Cassiopea frondosa*, respectively. Trench and Blank (1987) also described three new species of zooxanthellae, in addition to *S. microadriaticum*, that were associated with *C. xamachana*. Udy et al. (1993), however, pointed out that electron-dense chromosome-like structures within the nucleus are connected by a fine thread of DNA and may not be true chromosomes. To date, no additional studies on the chromosome number of zooxanthellae have been undertaken. Even though the electron-dense structures in the nucleus are not true chromosomes and are only parts of chromosomes, the nuclear morphology at the electron microscopy level differed greatly among zooxanthellae, likely justifying the conclusion that they belong to different species.

Since the early 1990s, zooxanthellae have been classified based on the sequence or restriction fragment length polymorphism (RFLP) patterns of the 18S rRNA gene (Rowan and Powers 1991a, b). Zooxanthellae are currently classified into nine clades based on nuclear and chloroplast ribosomal DNA (rDNA; Pochon and Gates 2010). Each clade corresponds to one branch of the phylogenetic tree of zooxanthellae, but each may consist of many species (LaJeunesse 2004; Correa and Baker 2009; Sampayo et al. 2009; LaJeunesse and Thornhill 2011). Various methods of zooxanthella classification or identification have been attempted. Initially, RFLP patterns of the 18S or 28S rRNA gene (rDNA) were used. Subsequently, zooxanthellae were further divided into subclades based on denaturing gradient gel electrophoresis (DGGE) or single-strand conformation polymorphism (SSCP) analysis and sequencing of the ITS2 or ITS1 region of rDNA.

The phylogenetic trees of eight representative *Symbiodinium* ITS2 types of clade C based on nuclear genes were congruent with those based on organelle genes, indicating that no genetic recombination occurred among these types (Sampayo et al. 2009). At least some ITS2 types may represent *Symbiodinium* species (Sampayo et al. 2009). If some ITS2 sequence variants are considered to potentially represent intraspecific sequence variation, some ITS2 clusters may correspond to *Symbiodinium* species (Correa and Baker 2009). The development and utilization of multiple, independent markers for *Symbiodinium* phylogenetic analysis have been attempted, and the use of each marker has been evaluated using clade or subclade level phylogenetic analysis (Pochon et al. 2012).

1.2.1.2 Symbiont Composition Within a Coral Host

The sensitivity of SSCP to detect minor symbiont components has been reported to be about 10 % (van Oppen and Gates 2007). Fragments with different nucleotide sequences can form a single band in DGGE analysis, resulting in underestimation of the diversity of zooxanthellae (van Oppen and Gates 2007; Apprill and Gates 2007). One method to describe the diversity of symbionts is to sequence a large number of clones of the ITS1 or ITS2 region for each symbiont DNA sample. Examining a greater number of clones may be necessary to detect the less abundant types (Apprill and Gates 2007). However, whether the minor sequence reflects the presence of minor symbiont components or of intragenomic variation of the ITS sequence is uncertain. Because rDNA has multiple copies, multiple sequences will be obtained if intragenomic variation in rDNA sequences exists (van Oppen and Gates 2007). Sampayo et al. (2009) suggested that the most commonly cloned sequence consistently matched the sequence of the dominant band(s) in a polymerase chain reaction (PCR)–DGGE pattern. However, they pointed out that the frequent recovery of pseudogenes, rare functional variants, and artifact sequences outweigh the potential of sequencing cloned sequences for identifying low-abundance components (Sampayo et al. 2009).

At present, no ideal method has been established to estimate the symbiont composition within a host. Mieog et al. (2007) analyzed the ITS1 region of rDNA using real-time PCR and were able to detect minor components at more than 100 times higher sensitivity compared to conventional techniques such as DDGE, SSCP, and RFLP analyses. In four coral species, *A. millepora*, *A. tenuis*, *Stylophora pistillata*, and *Turbinaria reniformis*, individual colonies are thought to harbor a single type of symbiont. However, Mieog et al. (2007) documented that 78 % of colonies of the above four species contained cryptic (minor) clade types in addition to the major type. This method has difficulty in quantifying the actual ratio of symbiont types because each clade of symbionts may contain a different number of rDNA copies and hence a different number of ITS1 regions. Mieog et al. (2009) used an intron of the actin gene, which has one to several copies in the genome, to successfully quantify the relative amount of clade C and D symbionts in coral colonies. In this method, correction for the number of actin gene copies was necessary (one and seven copies in clades C and D, respectively). Using real-time PCR and clade specific *Symbiodinium* primers, Correa et al. (2009) and Silverstein et al. (2012) detected minor clade D symbionts in coral hosts that had been considered to be associated with a single *Symbiodinium* clade other than clade D. The actual abundance of clade D relative to the other clade must be estimated after correction for differences in the number of copies of rDNA genes, DNA quality, and/or the efficiency of individual quantitative PCR reactions.

Using bacterial cloning and sequencing, Stat et al. (2011) reported that most *Montipora capitata* colonies from Kaneohe Bay, Hawaii, exhibited multiple ITS2 sequences of *Symbiodinium*. They suggested that the results may reflect actual *Symbiodinium* species diversity as well as intragenomic variants in a single *Symbiodinium* cell (Stat et al. 2011). LaJeunesse and Thornhill (2011) used the non-coding region of the plastid *psbA* minicircle (*psbA*^{ncr}), which is a rapidly evolving marker with low intragenomic variation, to analyze *Symbiodinium* diversity in colonies of *Montipora* spp. A single *psbA*^{ncr} haplotype was recovered in most samples through direct sequencing, and members of the same ITS2 type were phylogenetically differentiated from other ITS2 types based on *psbA*^{ncr} sequence divergence. The authors suggested that most colonies harbored one dominant *Symbiodinium* genotype (LaJeunesse and Thornhill 2011).

The fact that representatives of distinct rDNA clades are present within a single zooxanthella genome is quite intriguing. van Oppen and Gates (2007) described unpublished but interesting phenomena: van Oppen and Gates (2007) found that most ITS1 types within each *Symbiodinium* cell differed by two to eight substitutions and a few small indels, but that some ITS1 types from the same cell isolated from the corals *A. millepora* and *A. tenuis* were classified as either clade D or clade C. Similarly, Gates and coworkers (unpublished data) found that ITS sequences representative of clades B, C, and E were observed within single symbiont cells isolated from the sea anemone *Aiptasia pulchella*.

Recently, ITS2 sequences have been systematically obtained using next-generation sequencers in an attempt to determine symbiont composition within a host. Notably, ITS2 sequences from various clades have been obtained from a monoclonal culture of (or a single cell of) zooxanthella. This method may also be affected by intragenomic variation of the ITS2 sequence.

1.2.2 Diversity of Symbiotic Relationship Between Coral Host and Symbiotic Dinoflagellates

1.2.2.1 Changes in Symbiont Composition During and After Bleaching

Baker et al. (2004) reported that the composition of the *Symbiodinium* clade in corals changed after severe bleaching. In Kenya, the Persian Gulf, and Panama, where coral reefs have been severely affected by bleaching, many corals are associated with clade D *Symbiodinium*, whereas in Mauritius and the Red Sea, where corals were not affected by severe bleaching during 1997–2002, most corals are associated with clade C or clade C and A *Symbiodinium*, respectively. In Panama, the percentage of corals associated with clade D *Symbiodinium* increased after bleaching (Baker et al. 2004). In some corals, clade composition shifts when corals are transplanted from deep to shallow water (Baker 2001). In fact, photosynthesis rates in *P. damicornis* colonies hosting *Symbiodinium* D are higher than in corals with *Symbiodinium* C at high temperatures, suggesting that clade D symbionts are more stress tolerant than clade C symbionts (Rowan 2004). These observations suggest that corals may be able to adapt to global warming by switching their algal symbionts from stress-sensitive to stress-tolerant types as they recover from bleaching. This algal compositional shift can occur as a result of either the proliferation of endogenous symbionts that were minor components but are well suited to the new environment (shuffling) or the uptake of exogenous stress-tolerant symbionts (switching). However, debate exists as to whether the change in algal composition is temporary or stable (and thus leading to a new symbiotic combination).

Recent studies have shown that changes in algal composition caused by bleaching are not stable and that corals recovered from bleaching become associated with the original symbionts present before bleaching (Thornhill et al. 2005; Sampayo et al. 2008; Stat et al. 2009; LaJeunesse et al. 2010; McGinley et al. 2012). In most cases, the uptake of exogenous symbionts is transient, and recovery from bleaching is due to proliferation of the resident symbionts (Coffroth et al. 2010). Multiyear seasonal surveys have revealed that repetitively sampled corals from labeled colonies of six species from the Bahamas and the Florida Keys exhibited little or no change in their dominant symbiont type (Thornhill et al. 2005). In some colonies of *Montastrea annularis* and *Montastrea franksi* in Florida, clade D *Symbiodinium* was progressively replaced by symbionts typically found in these species, suggesting reversion to the original symbiont type from temporarily acquired or proliferated clade D *Symbiodinium* during bleaching. *S. pistillata* in the southern Great Barrier Reef (GBR) is associated with four types of clade C symbionts. Two of the four subclades (C79 and C35/a) are stress sensitive, and colonies associated with C79 or C35/a were bleached during a 2006 bleaching event. Although new types of symbionts were found during bleaching in these colonies, they became associated with the original symbionts C79 and C35/a if they recovered from bleaching (Sampayo et al. 2008). If the symbiont genotype was examined for randomly sampled colonies, the genotypic composition of symbionts in the population appeared to change: decreases in C79 and C35/a symbionts and increases in two stress-tolerant types. However, if the genotypic composition of symbionts were repeatedly examined on tagged colonies, the colonies associated with stress-sensitive types of symbionts died of bleaching. Thus, the differential susceptibilities of the holobionts were responsible for the apparent changes in the genotypic composition of symbionts in the population (Sampayo et al. 2008). Stat et al. (2009) investigated the symbiont composition of ten species, including five horizontal transmitters and five vertical transmitters, before and after a 2002 bleaching event. They found that the dominant *Symbiodinium* in the ten coral species did not change during or after the 2002 thermal stress event. Colonies of seven species were bleached, but they harbored the original subtypes when they recovered from bleaching. Colonies of the same species sometimes associated with different subtypes, but the symbiont type did not change after bleaching. In these corals, new types of symbionts sometimes appeared during bleaching, but this was temporary, and, when recovered from bleaching, the corals always associated with their original symbionts (Stat et al. 2009). Thus, shuffling and switching of symbiont type may not be as common in reef-building corals as previously thought (Baker 2001; Baker et al. 2004).

1.2.2.2 Flexibility of Symbiosis and Stress Tolerance

Even closely related species of coral may show different specificity in their symbiotic relationship. Sympatric colonies of *Pavona divaricata* and *Pavona decussata* exhibit marked differences in their specificity of symbiosis (Suwa et al. 2008). *P. divaricata* colonies can associate with clades C and D symbionts and often show seasonal variation in clade composition, whereas *P. decussata* colonies associate with clade C symbionts throughout the year. With its higher flexibility in symbiont type, *P. divaricata* was more resistant to seasonal environmental changes than *P. decussata*, which showed high fidelity to one symbiont clade (Suwa et al. 2008). These observations suggest that host flexibility in symbiont type as well as stress tolerance of symbionts play an important role in the stress tolerance of the holobiont. Silverstein et al. (2012) stated that the ability to associate with multiple symbiont clades is common in scleractinian corals. McGinley et al. (2012), however, reported that eastern Pacific *Pocillopora* spp. maintain long-term associations dominated by a specific *Symbiodinium* species (C1b-c or D1) despite a high prevalence of an alternate and compatible *Symbiodinium* species at low-abundance background levels. They suggested that potential competition by a second symbiont type is suppressed in the corals (McGinley et al. 2012).

Symbiotic flexibility is generally considered to enhance holobiont adaptation to environmental stress. However, *Acropora* and *Pocillopora* species, which show high symbiotic flexibility, are more stress sensitive than massive *Porites* corals, which generally show high fidelity to a single symbiont type (Putnam et al. 2012). The rate of environmental change due to global warming may be too rapid for corals with high symbiotic flexibility to adapt to the new environment via the acquisition of new symbionts or the shuffling of existing symbiont populations (Putnam et al. 2012). Long-lived corals such as the massive *Porites* may possess other mechanisms to adapt to environmental changes (see Sect. 1.3.4).

1.2.2.3 Symbiont Compositional Changes During Development

Planula larvae or juveniles (primary polyps and young colonies) may be associated with symbionts of diverse types, whereas adult colonies may be more specific to a symbiont type and generally associate with a single or a few types of symbionts. If *Symbiodinium*-free primary polyps of *A. tenuis* are transplanted to a reef, they acquire clade C and/or D symbionts, while adult colonies on the reef harbor predominantly clade C1 symbionts (Little et al. 2004). If juveniles of *A. tenuis* and *A. millepora* are transplanted to a reef where adult colonies were associated with clade C symbionts, they become primarily associated with clade D symbionts (Abrego et al. 2009a). Juveniles generally show low specificity or selectivity with symbiont type and associate mainly with symbionts that differ from those found in adult colonies but exhibit a high efficiency of infection (Abrego et al. 2009a).

If juveniles harboring clade C or D symbionts are transplanted to a reef, they do not acquire new types of symbionts from the environment. Abrego et al. (2009b) showed that juveniles of *A. tenuis* established association with adult homologous symbionts after about 3.5 years, whereas *A. millepora* juveniles did not change their *Symbiodinium* communities over the 3.5 years. These results suggest that these closely related species of *Acropora* differ in the timing of the onset of specificity for algal symbionts. One interesting hypothesis is that once coral juveniles associate with a certain type of symbiont, they develop an immunological tolerance to the type and lose the capacity to associate with other types of symbionts. Although it is unlikely that the innate immune system of corals exhibits immunological tolerance, whether the specificity for algal symbionts in adult colonies is simply the result of relative fitness/competition among symbionts or if it involves some host tolerance mechanisms should be examined further.

Vertical transmitters inherit symbionts from their mother colonies, yet whether they harbor only those symbionts inherited from their parent or if they acquire new symbionts from the environment like horizontal transmitters remains unclear. LaJeunesse et al. (2004) explored the symbiont composition of various coral hosts in GBR and Okinawa. They found that *Porites* and *Montipora* corals, both of which are vertical transmitters, associate with symbionts diversified from C15 and C21 subclades, respectively. Corals of *Porites* or *Montipora* in different regions associated with symbionts that had diverged on a region-specific basis from a certain ancestral type (LaJeunesse et al. 2004). These findings suggest that symbionts coevolved with these vertical transmitter corals in each region.

However, vertical transmitter corals may also acquire new types of symbionts from the environment, as in horizontal transmitter corals. Padilla-Gamiño et al. (2012) demonstrated that *Symbiodinium* ITS2 assemblages in the eggs of the brooding coral *M. capitata* are generally similar to their parents. However, in some cases, eggs contained *Symbiodinium* types that differed from those associated with their respective parent colonies. The difference in *Symbiodinium* ITS2 assemblages between eggs and their parent colonies may have been due to parental control (preferential transfer of *Symbiodinium* types to their eggs) or to stochastic infection related to the *Symbiodinium* composition of the fecund polyp, which may be associated with different *Symbiodinium* types due to microenvironmental differences within the colony. A third possibility is that eggs acquire *Symbiodinium* cells that are transiently present in the gastrovascular cavity of the parent polyp (Padilla-Gamiño et al. 2012). Similarly, juvenile colonies of the brooding coral *S. pistillata* may utilize both vertical and horizontal symbiont acquisition strategies (Byler et al. 2013). Shallow-water adults and planulae host clade A *Symbiodinium*, but if planulae harboring clade A *Symbiodinium* settle in deep water, these juveniles acquire clade C *Symbiodinium* from the environment and survive. Deep-water adults may harbor clade C *Symbiodinium* with potentially low levels of clade A *Symbiodinium*, but planulae only inherit clade C *Symbiodinium*. When these planulae settle in shallow water, they acquire clade A *Symbiodinium* and survive (Byler et al. 2013).

1.2.3 *Symbiodinium* Composition in the Environment

If nonsymbiotic juvenile corals or invertebrate hosts are transplanted to coral reefs, they acquire symbionts (Kinzie et al. 2001), suggesting that symbiotic dinoflagellates are available within the reef environment for invertebrate hosts. Zooxanthellae isolated from the sand bottom of coral reefs in Okinawa were all of clade A, and whether these clade A zooxanthellae are actually nonsymbiotic, free-living dinoflagellates or symbiotic ones remains unclear (Hirose et al. 2008b). Zooxanthellae isolated from the seawater of Hawaii and the Caribbean belonged to clades C and B, respectively (Manning and Gates 2008). While zooxanthellae isolated from seawater reflect the clades of symbionts associated with corals in that region, zooxanthellae associated with coral hosts (*P. cylindrica*, *M. capitata*, and *P. damicornis*) and those isolated from seawater in the same region were not identical at the subclade level (Pochon et al. 2010), suggesting that the incorporation of new types of symbionts from seawater is rare and switching of symbionts seldom occurs in these vertical transmitter corals. Moreover, zooxanthellae expelled from these corals likely do not survive for long in seawater.

Yamashita and Koike (2013) found that most of the environmental isolates of *Symbiodinium* from the Japanese coast (mainly from Okinawa, Nagasaki, and Kochi) form monophyletic subclades within the clade A lineage that are separate from a host-associated clade A population. 28S rDNA phylogeny and thecal-plate observations revealed that the environmental isolates were closely related to a "planktonic species," *Symbiodinium natans* described by Hansen and Daugbjerg (2009) and Yamashita and Koike (2013). However, environmental DNA clones were primarily nested within host-associated *Symbiodinium* groups in various clades. Yamashita and Koike (2013) suggested that the environmental *Symbiodinium* could be divided into two groups, one exclusively free living and the other transiently free living and possibly expelled from animal hosts.

1.3 Coral Bleaching and Cell Death

Tropical organisms generally show high tolerance to high temperature, strong light, and ultraviolet (UV) radiation. Coral–zooxanthella symbiotic systems are, however, sensitive to these environmental stresses and tend to suffer bleaching, i.e., disruption of the symbiotic relationship between the coral host and algal symbionts (Baird et al. 2009). Corals bleach if seawater temperatures increase beyond the summer maxima by 1–2 °C. If bleaching conditions last for a long period of time, bleached corals

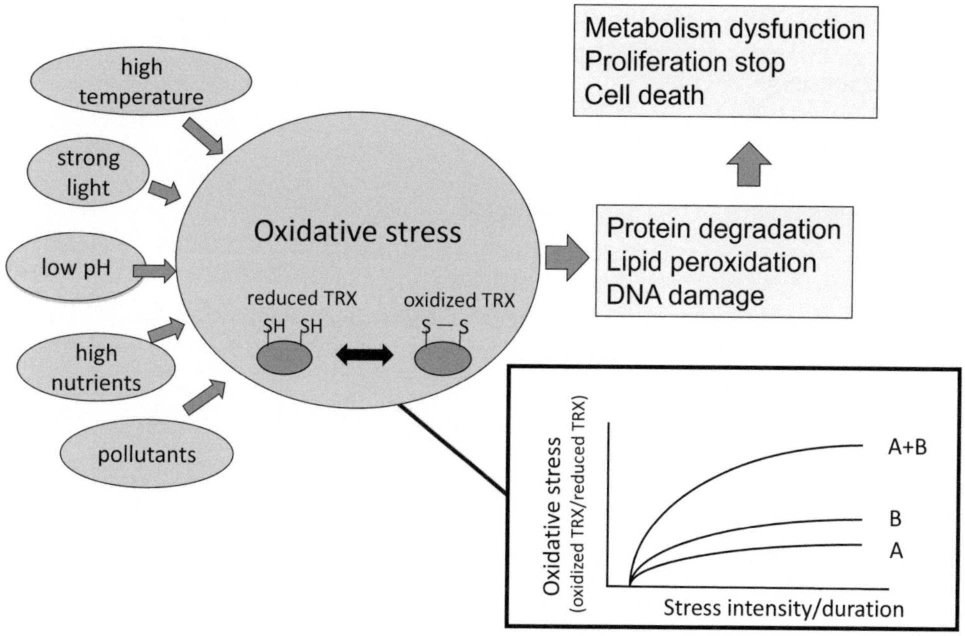

Fig. 1.9 Generalized model of stress response of coral cells. Various types of stresses cause oxidative stress within the cell (Partially based on Baird et al. 2009)

may die of nutrient deficiency. Coral species with a high capability of zooplankton feeding during bleaching and recovery may be more resilient to bleaching (Grottoli et al. 2006). However, the possibility exists that the presence of photodamaged symbionts in the tissue becomes a burden for host corals, as photodamaged symbionts become a source of ROS under light conditions. The resulting oxidative stress in the coral host cell leads to protein degradation, lipid peroxidation, and DNA damage. The oxidative damage further results in dysfunction of cellular metabolism, the arrest of proliferation, and ultimately cell death (Fig. 1.9; Baird et al. 2009). Thus, understanding the role of algal symbionts in the bleaching process of the host corals is indispensable.

1.3.1 Zooxanthellae Become a Burden for Host Coral Under Stressful Conditions

1.3.1.1 Experiments with Coral Cell Aggregates (Tissue Balls)

Dissociated coral cells or tissue isolates have been used as substitutes for coral colonies or planula larvae to examine the response of corals to environmental stress or toxic substances (Kopecky and Ostrander 1999; Domart-Coulon et al. 2004; Downs et al. 2010). If dissociated coral cells are kept in a Petri dish, they aggregate to form a ball-like structure (tissue ball), which rotates by ciliary movement. If such cell aggregates or tissue balls are exposed to high-temperature stress, they cease rotation and begin to disintegrate, dispersing cells from the tissue balls (Nesa and Hidaka 2009a). If this stage is considered the death of tissue balls, the survival time of tissue balls can be compared under different conditions. Tissue balls die more quickly under increased temperature (31 °C) than under normal temperature (25 °C; Fig. 1.10). Under high temperatures, Nesa and Hidaka (2009a) observed a significant negative correlation between the survival time of tissue balls and their zooxanthella density (Fig. 1.11). No such negative correlation was found under normal temperatures. These results suggest that tissue balls with higher densities of symbionts die more quickly under thermal stress than those with lower algal densities and that symbionts become a burden for host cells under such conditions. In a comet assay using tissue balls, Nesa and Hidaka (2009b) documented that host cells suffer DNA damage under thermal stress, but that antioxidants such as ascorbic acid and catalase or mannitol ameliorate the DNA damage of host cells (Fig. 1.12).

1.3.1.2 Experiments with Planula Larvae

Acropora species are horizontal transmitters and spawn zooxanthella-free eggs. The larvae of *Acropora* corals generally do not contain zooxanthellae, but inoculating larvae with zooxanthellae is possible (e.g., Harii et al. 2009). If such inoculated larvae and zooxanthella-free larvae are exposed to thermal stress (32 °C) for 3 days, zooxanthellate larvae exhibit higher rates of mortality, higher activity of superoxide dismutase (SOD), and increased contents of malondialdehyde, an indicator of lipid peroxidation, than zooxanthella-free larvae (Fig. 1.13; Yakovleva et al. 2009). If zooxanthellate and zooxanthella-free larvae of *A. tenuis*

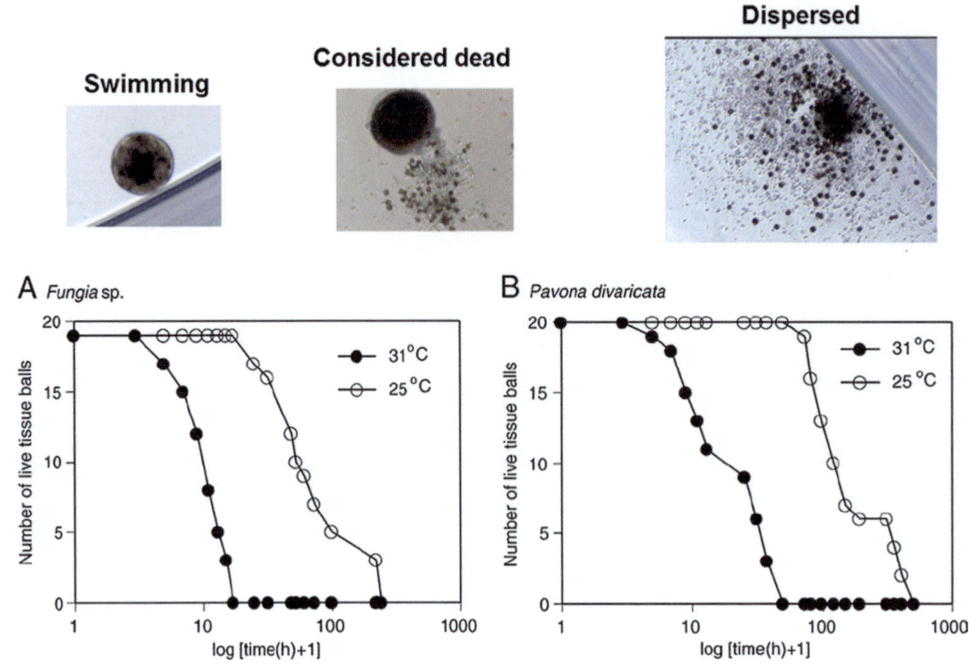

Fig. 1.10 Survivorship of coral cell aggregates (tissue balls) at different temperatures. Tissue balls died quickly at high temperature compared to normal temperature (From Nesa and Hidaka 2009a)

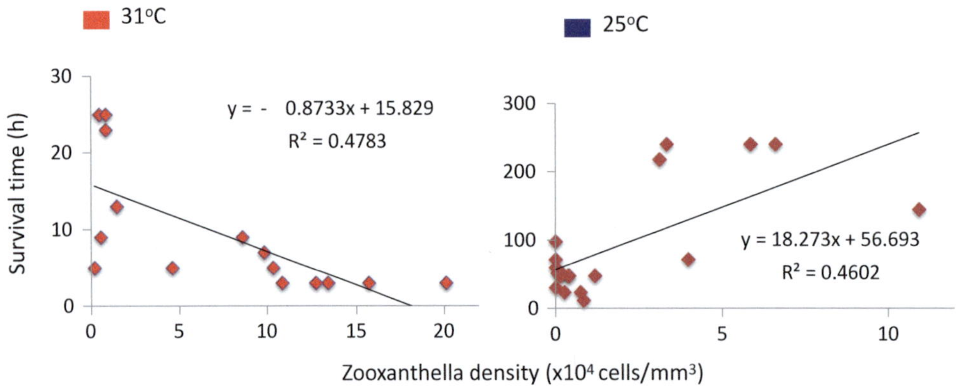

Fig. 1.11 The relationship between zooxanthella density of tissue balls and survival time. Negative correlation was found between survival time and zooxanthella density under thermal stress (From Nesa and Hidaka 2009a)

are exposed to natural sunlight during early summer in Okinawa, the zooxanthellate larvae suffer remarkable DNA damage, whereas zooxanthella-free larvae do not (Nesa et al. 2012). These observations suggest that the presence of zooxanthellae becomes a burden for planulae under thermal or light stress, as symbionts become a source of ROS under stressful conditions.

1.3.1.3 Adult Colonies

In adult corals, damage to the photosynthetic machinery of zooxanthellae caused by environmental stress leads to the generation of ROS, which in turn causes the apoptotic death of host cells (Weis 2008). The antioxidant activity of gastrodermal extract from the symbiotic sea anemone *Anemonia viridis* increased after elevated temperature stress, and this reaction was followed by the induction of caspase-like activity in the host gastrodermal cells (Richier et al. 2006). Caspases are cysteine aspartate-specific proteases that serve as initiators or effectors of apoptosis. Unbleached *S. pistillata* fragments produce twice more superoxide anion radical (O_2^-) in the light than in the dark, whereas light enhancement of O_2^- production was not observed in bleached fragments (Saragosti et al. 2010). These observations suggest that zooxanthella-containing corals tend to build up ROS within their tissue and are thus more susceptible to bleaching. When the sea

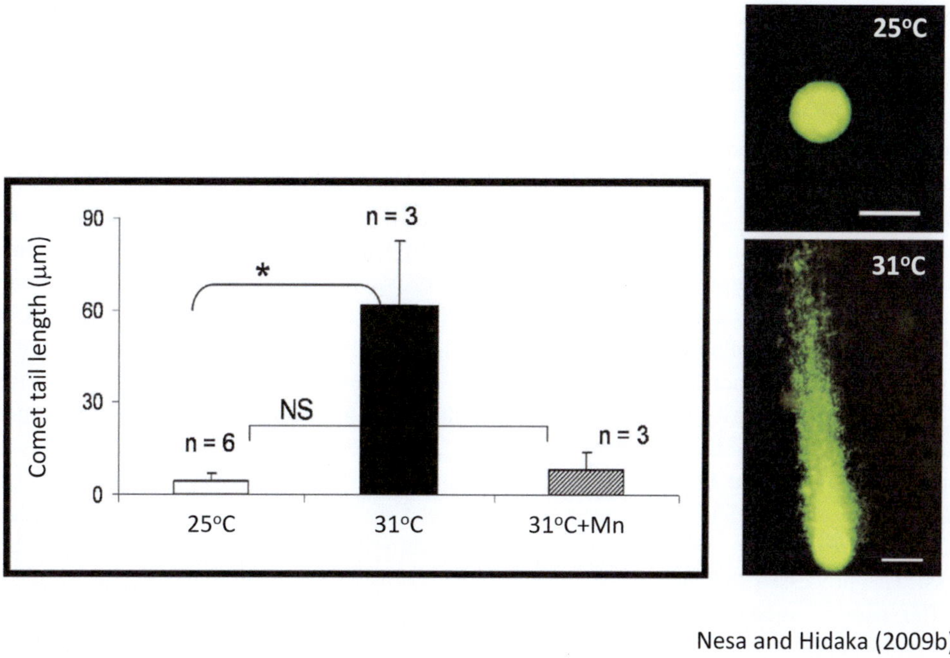

Fig. 1.12 Comet assay of tissue balls maintained at normal and high temperatures. Host cells suffered DNA damage under thermal stress. This damage was ameliorated by an exogenous antioxidant, mannitol (From Nesa and Hidaka 2009b)

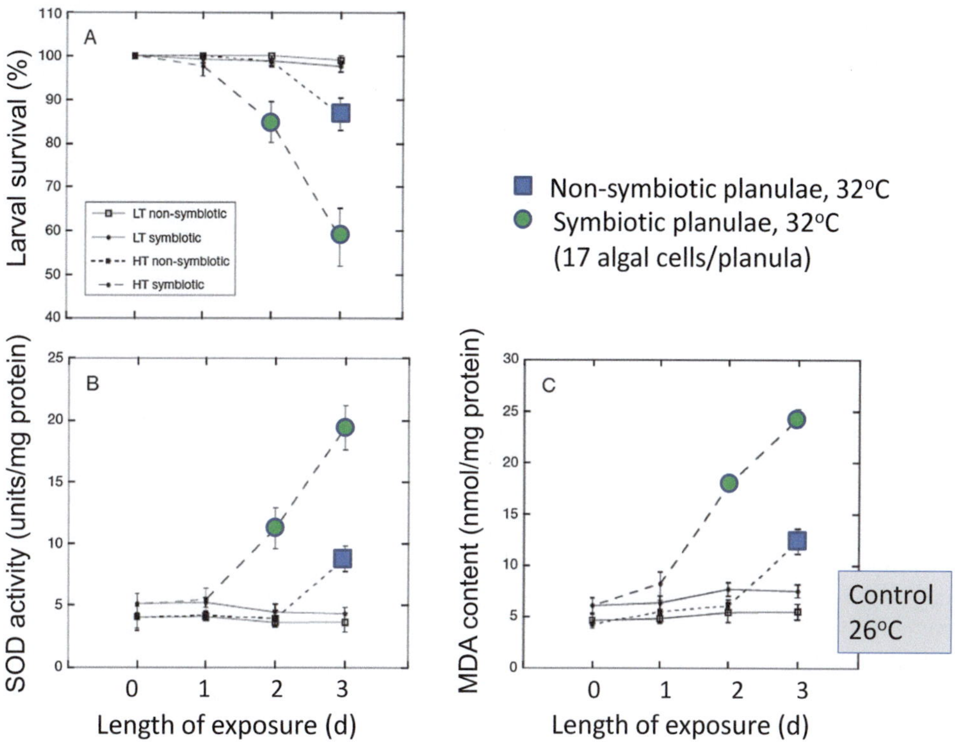

Fig. 1.13 Survivorship experiment with *Acropora intermedia* planulae. Symbiotic *A. intermedia* planulae had lower survival, higher activities of antioxidant enzyme, and higher levels of oxidative cellular damage (From Yakovleva et al. 2009)

anemone *Aiptasia* sp. or the coral *Acropora aspera* were exposed to thermal stress, apoptosis of gastrodermal cells was initially observed, but as the percentage of apoptotic cells reached a peak, the loss of symbionts or bleaching occurred (Dunn et al. 2002; Ainsworth et al. 2008). Subsequently, host gastrodermal cells disintegrated further, and degraded zooxanthellae were observed within the gastrodermis.

In naturally bleached corals, swelled zooxanthellae with many vacuoles are observed (Brown et al. 1995; Mise and Hidaka 2003). Occasionally, these zooxanthellae appear as if they have just ruptured. Franklin et al. (2004) reported that symbiotic dinoflagellates of the coral *S. pistillata* exposed to artificial bleaching conditions exhibited vacuolization, loss of chlorophyll, and increases in accumulation bodies. In the same experiment, the abundances of Sytox-green-stained algae and 2′,7′-dichlorodihydrofluorescein diacetate (H_2DCFDA)-stained algae increased, indicating compromised algal membrane integrity and cell death and increased oxidative stress, respectively. Strychar and Sammarco (2008) also reported that both apoptotic death and necrotic death of zooxanthellae are observed in *Acropora hyacinthus* and *Porites solida* during high-temperature stress. The processes that determine whether zooxanthellae suffer necrotic death or apoptotic death in corals are still not understood.

Chronic stress can cause the expulsion of degraded and photosynthetically incompetent zooxanthellae (Franklin et al. 2004; Strychar et al. 2004), whereas acute laboratory stress can also lead to the expulsion of healthy-looking and photosynthetically competent algal cells (Ralph et al. 2001; Bhagooli and Hidaka 2004a). Stress conditions, such as the intensity and duration of bleaching conditions, as well as host species identity, may affect the state of expelled zooxanthellae (Franklin et al. 2004; Strychar et al. 2004; Ralph et al. 2005; Hill and Ralph 2007).

1.3.2 Target of Thermal Stress and Determinant of Stress Sensitivity

1.3.2.1 Stress Sensitivity of Algal Symbionts

Recent studies support the idea that the stress sensitivity of the coral–zooxanthella symbiotic system is determined mainly by the algal partner. Corals associated with symbionts sensitive to elevated temperatures may die of bleaching under thermal stress, whereas those associated with stress-tolerant symbionts may survive. This process results in apparent shifts in the symbiont composition of a coral population (Sampayo et al. 2008).

1.3.2.2 D1 Protein Repair System as a Determinant of Stress Sensitivity

Warner et al. (1999) documented that the D1 protein repair system of algal symbionts is sensitive to high-temperature stress. D1 protein is a component of the photosystem II (PSII) reaction center; hence, its function is directly linked to the photochemical efficiency of PSII. The sensitivity of the D1 protein repair system to thermal stress is significantly correlated with the bleaching susceptibility of host corals (Takahashi et al. 2004). Because the degradation of D1 protein proceeds at a similar rate under thermal/light stress among corals with different stress susceptibilities, the D1 protein repair system is considered a determinant of the susceptibility of corals to bleaching (Takahashi et al. 2004). Differential sensitivities of de novo synthesis of thylakoid membrane proteins to high temperature corresponded to the degree of photodamage (decrease in PSII photochemical efficiency, F_v/F_m) among several cultured strains (Takahashi et al. 2009).

Because photochemical efficiency is closely linked to the intactness of D1 protein, one can easily understand that F_v/F_m and the functional integrity of D1 protein are strongly correlated. The functional integrity of D1 protein is determined by the balance between its degradation and repair. Therefore, it is important to investigate how the balance between the degradation and repair of D1 protein changes under thermal stress in various strains of zooxanthellae.

Decreases in F_v/F_m precede the bleaching of corals. However, whether this decrease in F_v/F_m causes coral bleaching or is a result of an ultimate cause of coral bleaching, such as ROS production by damaged symbionts, remains unclear. It is also not clear whether the breakdown of PSII leads to enhanced ROS production due to the Mehler reaction on the reducing side of PSI (Lesser 2011) or to failure of absorption of light energy, hence decreasing ROS production. If high-energy electrons produced in PSII cannot be consumed by the Calvin–Benson cycle, the extra electrons may react with ambient oxygen molecules to generate ROS. This situation occurs if some enzymes in the Calvin–Benson cycle are inactivated by thermal stress (Jones et al. 1998) and if thylakoid membrane integrity is damaged by increased temperature, leading to proton leakage and failure of ATP synthesis (Tchernov et al. 2004). Thus, thermal stress decreases the threshold of photoinhibition (Bhagooli and Hidaka 2004b), while D1 degradation results in the decreased production of high-energy electrons, thus serving as a safety valve under thermal stress. Yakovleva and Hidaka (2004) suggested that a delayed recovery of the electron transport rate compared to that of PSII photodamage also leads to ROS production and may therefore function as one determinant of varying susceptibilities to stress among coral species.

1.3.2.3 Role of the Host in Bleaching Susceptibility

If coral bleaching is the process by which ROS generation in algal symbionts leads to apoptotic death of host cells, many factors other than algal sensitivity to thermal stress may play important roles in determining the bleaching susceptibility of the coral. The thermal tolerance of host zooxanthellae does not always correspond to the thermal tolerance of isolated zooxanthellae, indicating that the host also plays a critical role in determining the stress susceptibility of the coral–zooxanthella symbiotic system (Bhagooli and Hidaka 2003). The antioxidant system of host corals as well as the photoprotection of algal symbionts provided by the host may affect the bleaching susceptibility of the holobiont.

Dunn et al. (2007) demonstrated that if both apoptosis and autophagy are simultaneously blocked using inhibitors (the caspase inhibitor Z-Val-Ala-Asp-fluoromethylketone (ZVAD-fmk) and 3-methyladenine and wortmannin, both of which are apoptosis inducers and autophagy inhibitors), bleaching is significantly diminished in the sea anemone *Aiptasia pallida* exposed to heat stress. Tchernov et al. (2011) reported that bleaching-sensitive *S. pistillata* exhibited typical signs of apoptosis after 1 week of exposure to an elevated temperature (32 °C), whereas bleaching-tolerant *S. hystrix* retained morphological integrity after 6 weeks of exposure. Caspase activity increased more than sixfold in *S. pistillata* but decreased by tenfold during the same time period in *S. hystrix*. These authors also found that the caspase inhibitor ZVAD-fmk prevented bleaching and apoptosis in both *P. damicornis* and *M. capitata* after 72 h of exposure to elevated temperature under full sunlight (Tchernov et al. 2011). If the caspase cascade were arrested at an early stage, the apoptotic response did not occur and the host survived the thermal stress, even though it sometimes suffered bleaching (Tchernov et al. 2011). The fate of coral host cells may be controlled by the balance between pro- and anti-apoptotic gene expression (Ainsworth et al. 2011; Pernice et al. 2011). Upregulation of the anti-apoptotic gene *Bcl-2* during thermal stress is followed by a delayed decrease in apoptotic activity (Pernice et al. 2011). Several studies have shown that nitric oxide (NO) plays an important role in the induction of apoptotic cell death during bleaching (Perez and Weis 2006; Bouchard and Yamasaki 2008).

Although algal symbionts become a burden for coral larvae, zooxanthella-containing larvae of the vertical transmitter coral *P. damicornis* are highly tolerant to thermal stress compared to zooxanthella-free larvae of the horizontal transmitter coral *A. tenuis* (Haryanti et al. 2015; Haryanti and Hidaka 2015). Zooxanthellate eggs of *M. capitata* contain higher Mn–SOD content than adult colonies (Padilla-Gamiño et al. 2013). Studies of the stress defense mechanisms of zooxanthella-containing larvae may provide insight into the mechanisms by which some corals exhibit high stress tolerance.

1.3.3 Stress Defense Mechanisms in Coral–Zooxanthella Symbiotic Systems

The symbiosis between corals and zooxanthellae evolved more than 200 million years ago during the Triassic (Stanley and Swart 1995; Stanley 2006). Considering the long history of coral–zooxanthella symbiosis, one must wonder why the coral–zooxanthella symbiotic system is so sensitive to environmental stress. The symbiotic system would be expected to evolve a defense system against environmental stresses and to be adapted to long-term environmental changes. Yet, recent global climate changes are likely occurring at a speed beyond the adaptation capacity of the coral–zooxanthella symbiotic system. To predict the response of coral reef ecosystems to global climate changes, understanding the adaptation mechanisms of the coral–zooxanthella symbiotic system to environmental changes is crucial.

Here, we consider possible collaborative defense systems of corals and symbiotic algae against environmental stresses. Corals contain the UV-absorbing substances mycosporine-like amino acids (MAAs), which function as sunscreen for algal symbionts, while some also function as antioxidants (Yakovleva et al. 2004). Thus, corals protect algal symbionts against UV radiation. Historically, animals including corals were thought to be unable to synthesize MAAs, as animals generally lack the shikimate pathway. Corals were once considered to take up MAAs from their diet or from algal endosymbionts and to utilize the incorporated MAAs directly or after modification. However, recent studies have reported that pentose metabolites serve as a source of MAA synthesis and that the primary MAA shinorine can be synthesized via a four-step pathway encoded by a gene cluster (Balskus and Walsh 2010; reviewed by Rosic 2012). A gene cluster homologous to the shinorine biosynthetic pathway can be found in various organisms including the sea anemone *Nematostella vectensis* (Balskus and Walsh 2010), the coral *A. digitifera* (Shinzato et al. 2011), and various strains of *Symbiodinium* (Rosic 2012). These findings indicate that coral hosts developed the ability to protect zooxanthellae from UV radiation damage.

Green fluorescent proteins (GFPs) in corals reportedly protect algal symbiont PSII from photoinhibition (Salih et al. 2000). Strong blue light causes photoinhibition (decrease in F_v/F_m) to a greater extent than green or red light (Oguchi et al. 2011). Because GFP absorbs blue light and emits green light, it is considered to have a photoprotective function, especially against blue light. Ten GFP-like protein genes have been found in the *A. digitifera* genome, and these genes code for GFP, CFP (cyan), RFP (red), and nonfluorescent chromoproteins (Shinzato et al. 2012). Diversification of GFP-like proteins in shallow-water corals may indicate a photoprotective function of the fluorescent protein in corals. GFP-like proteins also function as antioxidants (Bou-Abdallah et al. 2006; Palmer et al. 2009), again highlighting the presence of host protection of algal symbionts in the coral–zooxanthella symbiotic system. Notably, GFP in corals may have functions other than photoprotection, and the function of

GFP may differ among developmental stages or among species (Nakaema and Hidaka 2015a, b; Haryanti and Hidaka submitted).

Yuyama et al. (2010) found that host sulfate transporter gene expression increased when symbiosis with zooxanthellae was established in juveniles of *A. tenuis*. One possibility is that the incorporation of sulfate by the host is enhanced by the presence of symbionts and that zooxanthellae assimilate the incorporated sulfur and synthesize sulfur-containing amino acids such as cysteine and methionine. If sulfur-containing amino acids such as cysteine are transported to the host, they can be used to synthesize glutathione and thioredoxin (TRX), which play important roles in the intracellular antioxidant system and the transduction of intracellular oxidative signals. Intriguingly, the coral–zooxanthella symbiotic system may have developed a collaborative defense system against oxidative stress via translocation of sulfur and sulfur-containing amino acids between the host and symbiont. This hypothesis should be explored in studies of whether thermal stress increases sulfate transporter gene expression of host coral and whether knockdown of the sulfate transporter gene in the host decreases the stress tolerance of the coral–zooxanthella symbiotic system.

Acropora species lack a key enzyme necessary for cysteine synthesis, and these corals are highly sensitive to bleaching stress as they become dependent on symbionts for a supply of cysteine (Shinzato et al. 2011). In *Acropora* species, the interdependent defense system against oxidative stress likely functions under normal conditions but may collapse if the environment changes beyond the limit of tolerance of algal partners.

TRX contains thiol residues and regulates the oxidization/reduction of thiol residues of the target enzymes, thus regulating the activity of these enzymes (reviewed by Ahsan et al. 2010). TRX also plays an important role in the removal of hydrogen peroxide in conjunction with peroxiredoxin (TRX-dependent hydrogen peroxide reductase). Reduced TRX increases the DNA-binding capacity of various transcription factors, and it also binds to ASK1 (apoptosis-stimulating kinase 1) to inhibit apoptosis. One potential hypothesis is that zooxanthellae provide coral hosts with sulfur-containing amino acids and increase the stress tolerance of the host via production of TRX in host cells.

Various stresses commonly increase intracellular oxidative stress, although they may also induce stress-specific responses. Intracellular oxidative stress may serve as a good indicator of the early phases of stress responses to various types of stresses. If parameters of intracellular oxidative stress can be measured, we will be able to analyze the relationship between the stress response and the intensity or duration of the stress and further analyze whether different stresses act synergistically upon corals. Because the ratio of reduced TRX to oxidized TRX changes depending on intracellular oxidative stress levels, TRX functions as a signal transducer of oxidative stress. If the ratio of reduced TRX to oxidized TRX can be measured in coral tissues, it could serve as a useful parameter for intracellular oxidative stress. In humans, in response to oxidative stress, TRX becomes oxidized and released from the cells to the blood; thus, the TRX level in blood can be used as an indicator of stress levels (reviewed by Ahsan et al. 2010).

1.3.4 Adaptation of the Coral–Zooxanthella Symbiotic System to Environmental Changes

Historic convention implies that corals associated with symbionts that are sensitive to high-temperature stress may suffer bleaching, but when corals recover from bleaching, they may replace stress-sensitive symbionts with stress-tolerant symbionts (switching), or minor types of symbionts adapted to the new environment will become prevalent within the colony (shuffling) when the environment changes. Thus, corals would adapt to new environments such as warmer climates by shifting algal partners, and bleaching might be an adaptive process by which coral hosts survive under new environmental conditions (the adaptive bleaching hypothesis, ABH; Buddemeier and Fautin 1993). Contrary to the ABH, recent studies have indicated that many coral species are associated with specific types of symbionts and that the original algal composition is maintained after bleaching, even when temporary changes in algal composition occur during bleaching (Stat et al. 2009).

Zooxanthellae within cnidarian hosts are considered to be haploid cells proliferating asexually, although the genetic diversity of zooxanthella populations within octocoral colonies suggests the possibility of sexual recombination (Santos and Coffroth 2003). However, to date, no sexual reproduction has been reported for zooxanthellae in corals. If zooxanthellae only proliferate through asexual reproduction (mitotic cell division), they are clonal. However, considering a doubling time of several days and the huge number of algal cells within a colony, zooxanthellae cells may accumulate various mutations. Zooxanthella cells within a host colony could potentially belong to the same subclade or type and still be genetically different from one another, with or without mutations in some genes. In turn, these zooxanthella cells may exhibit different susceptibilities to environmental changes. When environmental changes occur, zooxanthella cells that are adaptable to the new environment may proliferate and become prevalent within the coral colony. This process is similar to somatic mutation in multicellular organisms, and each zooxanthella cell can be considered a stem cell (Correa and Baker 2011; van Oppen et al. 2011). This hypothesis is a new version of ABH or the shuffling of zooxanthella cells that are genetically different but belong to the same type or subclade.

Howells et al. (2012) demonstrated that generalist *Symbiodinium*-type (C1) populations associated with *A. tenuis* from two different thermal environments exhibited different thermal tolerances. *Symbiodinium* types can adapt to local environments, and this adaptation shapes the fitness of coral hosts. An interesting avenue of study would be to examine how quickly zooxanthellae accumulate mutations in genes involved in stress tolerance and how many such mutations are present within coral colonies. During ratchet experiment cycles, Huertas et al. (2011) reported that the number of generations required to grow under increased temperature (30 °C) was 55–70 for two *Symbiodinium* strains derived from corals. The ancestral strains maintained at 22 °C were unable to grow at 30 °C, but derived strains grew at 30 °C after 55–70 generations of the ratchet experiment. However, the derived strains were unable to grow at 35 °C and collapsed during the experiment.

Whether corals possess stem cells like hydrozoan I-cells is not well understood. However, epithelial cells of many cnidarians potentially have the capacity of self-renewal and transdifferentiation to various cell types (Gold and Jacobs 2013). If corals accumulate such stemlike cells with various mutations, certain stemlike cells that have acquired mutations adaptable to new environments may proliferate and replace somatic cells within the colony. If this is the case, aged colonies such as huge massive *Porites* colonies have a higher capacity to adapt to environmental changes compared to corals with determinate growth and short life spans, such as *Acropora* and *Pocillopora* colonies.

1.4 Stress Response of Coral–Zooxanthella Symbiotic Complex at the Gene Level

If we are to predict how coral reefs respond to local pollution and global climate changes, one must understand the responses of the coral–zooxanthella symbiotic system to environmental stress as well as the maintenance mechanisms of the symbiotic relationship between coral hosts and symbionts at the cellular and molecular levels. Several studies provide excellent reviews of the establishment and maintenance of the symbiotic relationship between cnidarian hosts and symbionts (Davy et al. 2012; Hill and Hill 2012). The responses of the coral–zooxanthella symbiotic system to environmental stress have been intensively studied during the past decade (Weis 2008; Roth 2014). Here, only several early-stage studies are discussed.

1.4.1 Stress-Specific Biomarkers

To protect and restore coral reefs, one must determine the types of stresses to which corals are exposed and whether they are exposed to single or combined stresses. The response of corals to environmental stress has been studied at both molecular and gene expression levels. A broad range of genes exhibit changes in expression under stress, and some genes appear to respond to specific types of stress, indicating the possibility of using such genes as stress-specific markers.

Desalvo et al. (2008) analyzed differences in gene expression between normal and bleached colonies of *Montastraea faveolata* using cDNA microarrays and then investigated differentially expressed genes in corals exposed to thermal stress for 9 days. Thermal stress and subsequent bleaching influenced the expression of genes with broad ranges of function, such as oxidative stress responses, calcium ion homeostasis, cytoskeletal functions, cell death, calcification, metabolism, protein synthesis, and heat shock proteins. Meyer et al. (2009) performed a transcriptome analysis of *A. millepora* planulae exposed to thermal stress using a next-generation sequencer 454. They found many differentially expressed genes, including those involved in the stress response, apoptosis, immune response, and protein folding. In addition to these, genes involved in metabolism, intracellular signal transduction, and transcription factors became available in the database (Meyer et al. 2009). Genome information for *A. digitifera* (Shinzato et al. 2011) established an infrastructure for studying the response of *Acropora* corals to environmental stress at the gene level.

Using reverse transcription (RT)-PCR, Smith-Keune and Dove (2008) observed that if colonies of *A. millepora* are exposed to high temperature (32 °C) for 6 h, the expression of the GFP homologue decreased. The decrease in GFP homologue expression occurred in advance to bleaching, indicating that GFP is a sensitive marker of thermal stress. Rodriguez-Lanetty et al. (2009) also reported that the expression of fluorescent protein (DsRed-type FP) decreased if planula larvae of *A. millepora* were exposed to high temperature (31 °C) stress for 3 h. They suggested that DsRed-type FP might be the same as the GFP homologue studied by Smith-Keune and Dove (2008). In larvae exposed to 3 h of high-temperature stress, the expression of heat shock protein increased, whereas mannose-binding C-type lectin was downregulated (Rodriguez-Lanetty et al. 2009). The authors suggested that lectin is involved in recognition of pathogens; hence, decreased expression of the lectin gene might be related to increased sensitivity to disease (decreased immunocompetency).

Using the HiCEP method, Yuyama et al. (2012) examined the gene expression of primary polyps of *A. tenuis* exposed to either high temperature (32 °C), organic tin, or photosynthetic inhibitor dichlorophenyl dimethylurea (DCMU). In the HiCEP method, restriction enzyme-digested cDNA fragments are used as a PCR template using fluorescence-labeled primers. The amount of each amplicon is analyzed using a sequencer. Yuyama et al. (2012) compared gene expression among different samples by comparing peak intensity among samples. They found 98 upregulated genes, of which 9 genes exhibited upregulation in response to all three stresses, 27 genes responded to two of three stresses, and 62 responded to a single type of stress. Only 7 of 98 genes, 5 coral and 2 zooxanthella genes, were annotated. One (oxidative stress-responsive protein) of the seven annotated genes responded to all three types of stress. Because most genes responded differently to different types of stress, these results suggest the possibility of identifying the type of stress affecting corals in the field using gene expression analysis.

1.4.2 Genes Involved in Symbiosis Between Cnidarians and Zooxanthellae

Understanding how the symbiosis between corals and zooxanthellae is maintained is crucial for determining the mechanisms of bleaching and breakdown of the symbiosis. Yuyama et al. (2010) examined the gene expression of primary polyps of *A. tenuis* before and after the establishment of symbiosis using HiCEP analysis. Genes involved in lipid metabolism, intracellular signal transduction, and membrane transport exhibited differential expression. In addition, the sulfate transporter gene was upregulated by the establishment of symbiosis.

Voolstra et al. (2009) inoculated planula larvae of *Acropora palmata* and *M. faveolata* with compatible and incompatible strains of symbionts and investigated differentially expressed genes using microarray analysis. They found that if planulae are inoculated with compatible symbionts, only a few genes exhibited changes in expression levels. In contrast, if planulae were inoculated with incompatible symbionts, many genes exhibited marked differences in expression levels 6 days after inoculation. Voolstra et al. (2009) suggested that regulation of apoptotic pathways and the immune system, via such processes as signal transduction via MAPK and the transcription factor NF-κB, is important for the establishment of symbiosis.

Sunagawa et al. (2009) analyzed 10,285 expressed sequence tags (ESTs) from the cDNA library of the sea anemone *A. pallida* to identify genes involved in symbiosis between cnidarian hosts and algal symbionts. The genes involved in symbiosis included genes of the oxidative stress response system, such those involved in the synthesis and oxidation/reduction of glutathione and enzymes that catalyze oxidation of thioredoxin.

Few studies, however, have examined gene expression changes of algal symbionts during the establishment of symbiosis. Bertucci et al. (2010) demonstrated that symbiont P-type H^+-ATPase expression was upregulated by the establishment of symbiosis. P-type H^+-ATPase functions as an H^+-pump that performs transmembrane transport of H^+ using energy from ATP hydrolysis. The pump is likely involved in lowering the internal pH of symbiosomes to increase the carbon dioxide (CO_2) concentration by accelerating the production of CO_2 from bicarbonate (HCO_3^-), thus providing the photosynthetic system with CO_2 (concentrating CO_2 for photosynthetic use). Many key questions regarding the cell biology of cnidarian–dinoflagellate symbiosis remain unresolved and await further research (Davy et al. 2012).

References

Abrego D, van Oppen MJH, Willis BL (2009a) Highly infectious symbiont dominates initial uptake in coral juveniles. Mol Ecol 18:3518–3531

Abrego D, van Oppen MJH, Willis BL (2009b) Onset of algal endosymbiont specificity varies among closely related species of *Acropora* corals during early ontogeny. Mol Ecol 18:3532–3543

Adams LM, Cumbo VR, Takabayashi M (2009) Exposure to sediment enhances primary acquisition of *Symbiodinium* by asymbiotic coral larvae. Mar Ecol Prog Ser 377:149–156

Ahsan MK, Nakamura H, Yodoi J (2010) Redox regulation by thioredoxin in cardiovascular diseases. In: Das DK (ed) Methods in redox signaling. Mary Ann Liebert, New York, pp 159–165

Ainsworth TD, Hoegh-Guldberg O, Heron SF, Skirving WJ, Leggat W (2008) Early cellular changes are indicators of pre-bleaching thermal stress in the coral host. J Exp Mar Biol Ecol 364:63–71

Ainsworth TD, Wasmund K, Ukani L, Seneca F, Yellowees D, Miller D, Leggat W (2011) Defining the tipping point. A complex cellular life/death balance in corals in response to stress. Sci Rep 1:160. doi:10.1038/srep00160

Apprill AM, Gates RD (2007) Recognizing diversity in coral symbiotic dinoflagellate communities. Mol Ecol 16:1127–1134

Ayre DJ, Resing JM (1986) Sexual and asexual production of planulae in reef corals. Mar Biol 90:187–190

Babcock RC, Heyward AJ (1986) Larval development of certain gamete-spawning scleractinian corals. Coral Reefs 5:111–116

Baird AH, Bhagooli R, Ralph PJ, Takahashi S (2009) Coral bleaching: the role of the host. Trends Ecol Evol 24:16–20

Baker AC (2001) Reef corals bleach to survive change. Nature 411:765–766

Baker AC, Starger CJ, McClanahan T, Glynn PW (2004) Corals' adaptive response to climate change. Nature 430:741

Balskus EP, Walsh CT (2010) The genetic and molecular basis for sunscreen biosynthesis in cyanobacteria. Science 329:1653–1656

Bertucci A, Tambutté E, Tambutté S, Aallemand D, Zoccola D (2010) Symbiosis-dependent gene expression in coral-dinoflagellate association: cloning and characterization of a P-type H^+-ATPase gene. Proc R Soc B 277:87–95

Bhagooli R, Hidaka M (2003) Comparison of stress susceptibility of in hospite and isolated zooxanthellae among five coral species. J Exp Mar Biol Ecol 291:181–197

Bhagooli R, Hidaka M (2004a) Release of zooxanthellae with intact photosynthetic activity by the coral *Galaxea fascicularis* in response to high temperature stress. Mar Biol 145:329–337

Bhagooli R, Hidaka M (2004b) Photoinhibition, bleaching susceptibility and mortality in two scleractinian corals, *Platygyra ryukyuensis Stylophora pistillata*, in response to thermal and light stresse. Comp Biochem Physiol Part A 137:547–555

Bouchard JN, Yamasaki H (2008) Heat stress stimulates nitric oxide production in *Symbiodinium microadriaticum*: a possible linkage between nitric oxide and the coral bleaching phenomenon. Plant Cell Physiol 49:641–652

Bou-Abdallah F, Chasteen ND, Lesser MP (2006) Quenching of superoxide radicals by green fluorescent protein. Biochim Biophys Acta 1760:1690–1695

Brown BE, LeTissier MDA, Bythell JC (1995) Mechanisms of bleaching deduced from histological studies of reef corals sampled during a natural bleaching event. Mar Biol 122:655–663

Buddemeier RW, Fautin DG (1993) Coral bleaching as an adaptive mechanism. BioScience 43:320–326

Byler KA, Carmi-Veal M, Fine M, Goulet TL (2013) Multiple symbiont acquisition strategies as an adaptive mechanism in the coral *Stylophora pistillata*. PLoS One 8(3):e59596

Coffroth MA, Poland DM, Petrou EL, Brazeau DA, Holmberg JC (2010) Environmental symbiont acquisition may not be the solution to warming seas for reef-building corals. PLoS One 5(10):e13258

Correa AMS, Baker AC (2009) Understanding diversity in coral-algal symbiosis: a cluster-based approach to interpreting fine-scale genetic variation in the genus *Symbiodinium*. Coral Reefs 28:81–93

Correa AMS, Baker AC (2011) Disaster taxa in microbially mediated metazoans: how endosymbionts and environmental catastrophes influence the adaptive capacity of reef corals. Glob Chang Biol 17:68–75

Correa AMS, McDonald MD, Baker AC (2009) Development of clade-specific *Symbiodinium* primers for quantitative PCR (qPCR) and their application to detect clade D symbionts in Caribbean corals. Mar Biol 156:2403–2411

Davy SK, Allemand D, Weis VM (2012) Cell biology of cnidarian-dinoflagellate symbiosis. Microbiol Mol Biol Rev 76:229–261

Desalvo MK, Voolstra CR, Sunagawa S, Schwarz JA, Stillman JH, Coffroth MA, Szmant AM, Medina M (2008) Differential gene expression during thermal stress and bleaching in the Caribbean coral *Montastraea faveolata*. Mol Ecol 17:3952–3971

Domart-Coulon I, Tambutté S, Tambutté E, Allemand D (2004) Short term viability of soft tissue detached from the skeleton of reef-building corals. J Exp Mar Biol Ecol 309:199–217

Downs CA, Fauth JE, Downs VD, Ostrander GK (2010) In vivo cell-toxicity screening as an alternative animal model for coral toxicology: effects of heat stress, sulfide, rotenone, cyanide, and cuprous oxide on cell viability and mitochondrial function. Ecotoxicology 19:171–184

Dunn SR, Bythell JC, Le Tissier MDA, Burnett WJ, Thomason JC (2002) Programmed cell death and cell necrosis activity during hyperthermic stress-induced bleaching of the symbiotic sea anemone *Aiptasia* sp. J Exp Mar Biol Ecol 272:29–53

Dunn SR, Schnitzler CE, Weis VM (2007) Apoptosis and autophagy as mechanisms of dinoflagellate symbiont release during cnidarian bleaching: every which way you lose. Proc R Soc B 274:3079–3085

Fautin DG, Mariscal RN (1991) Cnidaria: anthozoa. In: Microscopic anatomy of invertebrates, vol 2, Placozoa, Porifera, Cnidaria, and Ctenophora. Wiley-Liss, New York, pp 267–358

Franklin DJ, Hoegh-Guldberg O, Jones RJ, Berges JA (2004) Cell death and degeneration in the symbiotic dinoflagellates of the coral *Stylophora pistillata* during bleaching. Mar Ecol Prog Ser 272:117–130

Geller JB, Fitzgerald LJ, King CE (2005) Fission in sea anemones: integrative studies of life cycle evolution. Integr Comp Biol 45:615–622

Gold DA, Jacobs DK (2013) Stem cell dynamics in Cnidaria: are there unifying principles? Dev Genes Evol 223:53–66

Grottoli AG, Rodrigues LJ, Palardy JE (2006) Heterotrophic plasticity and resilience in bleached corals. Nature 440:1186–1189

Hansen G, Daugbjerg N (2009) *Symbiodinium natans* sp. nov.: a 'free-living' dinoflagellate from Tenerife (Northeast-Atlantic Ocean). J Phycol 45:251–263

Harii S, Yasuda N, Lodoriguez-Lanetty IT, Hidaka M (2009) Onset of symbiosis and distribution patterns of symbiotic dinoflagellates in the larvae of scleractinian corals. Mar Biol 156:1203–1212

Harii S, Yamamoto M, Hoegh-Guldberg O (2010) The relative contribution of dinoflagellate photosynthesis and stored lipids to the survivorship of symbiotic larvae of the reef-building corals. Mar Biol 157:1215–1224

Harrison PL (2011) Sexual reproduction of scleractinian corals. In: Dubinsky Z, Stambler N (eds) Coral reefs: an ecosystem in transition. Springer, Dordrecht, pp 59–85

Haryanti D, Hidaka M (2015) Temperature dependence of respiration in larvae and adult colonies of the corals *Acropora tenuis Pocillopora damicornis*. J Mar Sci Eng 3:509–519. doi:10.3390/jmse3030509

Haryanti D, Hidaka M (submitted) Developmental changes in the fluorescence intensity and distribution pattern of green fluorescent protein (GFP) in coral larvae and juveniles. Submitted

Haryanti D, Yasuda N, Harii S, Hidaka M (2015) High tolerance of symbiotic larvae of *Pocillopora damicornis* to thermal stress. Zool Stud 54:52. doi:10.1186/s40555-015-0134-7

Highsmith RC (1982) Reproduction by fragmentation in corals. Mar Ecol Prog Ser 7:207–226

Hill M, Hill A (2012) The magnesium inhibition and arrested phagosome hypotheses: new perspectives on the evolution and ecology of *Symbiodinium* symbioses. Biol Rev 87:804–821

Hill R, Ralph PJ (2007) Post-bleaching viability of expelled zooxanthellae from the scleractinian coral *Pocillopora damicornis*. Mar Ecol Prog Ser 352:137–144

Hirose M, Hidaka M (2006) Early development of zooxanthella-containing eggs of the corals *Porites cylindrica* and *Montipora digitata*: the endodermal localization of zooxanthellae. Zool Sci 23:873–881

Hirose M, Kinzie RA III, Hidaka M (2001) Timing and process of entry of zooxanthellae into oocytes of hermatypic corals. Coral Reefs 20:273–280

Hirose M, Yamamoto H, Nonaka M (2008a) Metamorphosis and acquisition of symbiotic algae in planula larvae and primary polyps of *Acropora* spp. Coral Reefs 27:247–254

Hirose M, Reimer JD, Hidaka M, Suda S (2008b) Phylogenetic analyses of potentially free-living *Symbiodinium* spp. isolated from coral reef sand in Okinawa, Japan. Mar Biol 155:105–112

Hoadley KD, Szmant AM, Pyott SJ (2011) Circadian clock gene expression in the coral *Favia fragum* over diel and lunar reproductive cycles. PLoS One 6:e19755

Howells EJ, Beltran VH, Larsen NW, Bay LK, Willis BL, van Oppen MJH (2012) Coral thermal tolerance shaped by local adaptation of photosymbionts. Nat Clim Chang 2:116–120

Huang H-J, Wang L-H, Chen W-NU, Fang L-S, Chen C-S (2008) Developmentally regulated localization of endosymbiotic dinoflagellates in different tissue layers of coral larvae. Coral Reefs 27:365–372

Huertas IE, Rouco M, López-Rodas V, Costas E (2011) Warming will affect phytoplankton differently: evidence through a mechanistic approach. Proc R Soc B 278:3534–3543

Hughes TP, Jackson JBC (1985) Population dynamics and life histories of foliaceous coral. Ecol Monogr 55:141–166

Jokiel PL, Bigger CH (1994) Aspects of histocompatibility and regeneration in the solitary reef coral *Fungia scutaria*. Biol Bull 186:72–80

Jones RJ, Hoegh-Gulberg O, Larkum AWD, Schreiber U (1998) Temperature-induced bleaching of corals begins with impairment of the

CO_2 fixation mechanism in zooxanthellae. Plant Cell Environ 21:1219–1230

Kerr AM, Baird AH, Hughes TP (2011) Correlated evolution of sex and reproductive mode in corals (Anthozoa: Scleractinia). Proc R Soc B 278:75–81

Kinzie RA III, Takayama M, Santos SR, Coffroth MA (2001) The adaptive bleaching hypothesis: experimental tests of critical assumptions. Biol Bull 200:51–58

Kojis BL (1986) Sexual reproduction in *Acropora* (*Isopora*) (Coelenterata: Scleractinia) I. *A. cuneata* and *A. palifera* on Heron Island reef, Great Barrier Reef. Mar Biol 91:291–309

Kopecky EJ, Ostrander GK (1999) Isolation and primary culture of viable multicellular endothelial isolates from hard corals. In Vitro Cell Dev Biol-Anim 35:616–624

Kramarsky-Winter E, Loya Y (1996) Regeneration versus budding in fungiid corals: a trade-off. Mar Ecol Prog Ser 134:179–185

LaJeunesse TC (2004) "Species" radiations of symbiotic dinoflagellates in the Atlantic and Indo-Pacific since the Miocene-Pliocene transition. Mol Biol Evol 22:570–581

LaJeunesse TC, Thornhill DJ (2011) Improved resolution of reef-coral endosymbiont (*Symbiodinium*) species diversity, ecology, and evolution through *psbA* non-coding region genotyping. PLoS One 6 (12):e29013

LaJeunesse TC, Bhagooli R, Hidaka M, deVantier L, Done T, Schmidt GW, Fitt WK, Hoegh-Guldberg O (2004) Shifts in relative dominance between closely related *Symbiodinium* spp. in coral reef host communities over environmental, latitudinal, and biogeographic gradients. Mar Ecol Prog Ser 284:147–161

LaJeunesse TC, Smith R, Walther M, Pinzón J, Pettay DT, McGinley M, Aschaffenburg M, Medina-Rosas P, Cupul-Magaña AL, Pérez AL, Reyes-Bonilla H, Warner ME (2010) Host-symbiont recombination versus natural selection in the response of coral-dinoflagellates symbiosis to environmental disturbance. Proc R Soc B 277:2925–2934

Lesser MP (2011) Coral bleaching: causes and mechanisms. In: Coral reefs: an ecosystem in transition. Springer, Dordrecht, pp 405–419

Levy O, Appelbaum L, Leggat W, Gothlif Y, Hayward DC, Miller DJ, Hoegh-Guldberg O (2007) Light-responsive cryptochromes from a simple multicellular animal, the coral *Acropora millepora*. Science 318:467–480

Little AF, van Oppen MJH, Willis BL (2004) Flexibility in algal endosymbioses shapes growth in reef corals. Science 304:1492–1494

Lough JM, Barnes DJ (1997) Several centuries of variation in skeletal extension, density and calcification in massive *Porites* colonies from the Great Barrier Reef: a proxy for seawater temperature and a background of variability against which to identify unnatural change. J Exp Mar Biol Ecol 211:29–67

Manning MM, Gates RD (2008) Diversity in populations of free-living *Symbiodinium* from a Caribbean and Pacific reef. Limnol Oceanogr 53:1853–1861

Marlow HQ, Martindale MQ (2007) Embryonic development in two species of scleractinian coral embryos: *Symbiodinium* localization and mode of gastrulation. Evol Dev 9:355–367

McGinley MP, Aschaffenburg MD, Pettay DT, Smith RT, LaJeunesse TC, Warner ME (2012) *Symbiodinium* spp. in colonies of eastern Pacific *Pocillopora* spp. are highly stable despite the prevalence of low-abundance background populations. Mar Ecol Prog Ser 462:1–7

Meyer E, Aglyamova GV, Wang S, Buchanan-Carter J, Abrego D, Colbourne JK, Willis BL, Matz MV (2009) Sequencing and *de novo* analysis of a coral larval transcriptome using 454 GSFlx. BMC Genomics 10:219

Mieog JC, van Oppen MJH, Cantin NE, Stam WT, Olsen JL (2007) Real-time PCR reveals a high incidence of *Symbiodinium* clade D at low levels in four scleractinian coral across the Great Barrier Reef: implication for symbiont shuffling. Coral Reefs 26:449–457

Mieog JC, van Oppen MJH, Berkelmans R, Stam WT, Olsen JL (2009) Quantification of algal endosymbionts (*Symbiodinium*) in coral tissue using real-time PCR. Mol Ecol Resour 9:74–82

Miller SW, Hayward DC, Bunch TA, Miller DJ, Ball EE, Bardwell VJ, Zarkower D, Brower DL (2003) A DM domain protein from a coral, *Acropora millepora*, homologous to proteins important for sex determination. Evol Dev 5:251–258

Miller DJ, Ball EE, Technau U (2005) Cnidarians and ancestral genetic complexity in the animal kingdom. Trends Genet 21:536–539

Mise T, Hidaka M (2003) Degradation of zooxanthellae in the coral *Acropora nasuta* during bleaching. Galaxea, JCRS 5:32–38

Nakaema S, Hidaka M (2015a) Fluorescent protein content and stress tolerance of two color morphs of the coral *Galaxea fascicularis*. Galaxea J Coral Reef Stud 17:1–11

Nakaema S, Hidaka M (2015b) GFP distribution and fluorescence intensity in *Galaxea fascicularis*: developmental changes and maternal effects. Platax 12:19–27

Nesa B, Hidaka M (2009a) High zooxanthella density shortens the survival time of coral cell aggregates under thermal stress. J Exp Mar Biol Ecol 368:81–87

Nesa B, Hidaka M (2009b) Thermal stress increases oxidative DNA damage in coral cell aggregates. In: Proceedings of 11th international coral reef symposium (Florida), pp 144–148

Nesa B, Baird AH, Harii S, Yakovleva I, Hidaka M (2012) Algal symbionts increase DNA damage in coral planulae exposed to sunlight. Zool Stud 51:12–17

Nilsson Sköld H, Obst M (2011) Potential for clonal animals in longevity and ageing studies. Biogerontology 12:387–396

Oguchi R, Terashima I, Kou J, Chow WS (2011) Operation of dual mechanisms that both lead to photoinactivation of photosystem II in leaves by visible light. Physiol Plant 142:47–55

Ojimi MC, Loya Y, Hidaka M (2012) Sperm of the solitary coral *Ctenactis echinata* exhibits a longer telomere than that of somatic tissues. Zool Stud 51:1475–1480

Padilla-Gamiño JL, Pochon X, Bird C, Concepcion GT, Gates RD (2012) From parent to gamete: vertical transmission of *Symbiodinium* (Dinophyceae) ITS2 sequence assemblages in the reef building coral *Montipora capitata*. PLoS One 7:e38440

Padilla-Gamiño JL, Bidigare RR, Barshis DJ, Alamaru A, Hédouin L, Hernández-Pech X, Kandel F, Leon Soon S, Roth MS, Rodrigues LJ, Grottoli AG, Portocarrero C, Wagenhauser SA, Buttler F, Gates RD (2013) Are all eggs created equal? A case study from the Hawaiian reef-building coral *Montipora capitata*. Coral Reefs. doi:10.1007/s00338-012-0957-1

Palmer CV, Chintan KM, Laura DM (2009) Coral fluorescent proteins as antioxidants. PLoS One 4:e7298. doi:10.1371/journal.pone.0007298

Perez S, Weis V (2006) Nitric oxide and cnidarian bleaching: an eviction notice mediates breakdown of a symbiosis. J Exp Biol 209:2804–2810

Permata DW, Hidaka M (2005) Ontogenetic changes in the capacity of the coral *Pocillopora damicornis* to originate branches. Zool Sci 22:1197–1203

Permata DW, Kinzie RA III, Hidaka M (2000) Histological studies on the origin of planulae of the coral *Pocillopora damicornis*. Mar Ecol Prog Ser 200:191–200

Pernice M, Dunn SR, Miard T, Dufour S, Dove S, Hoegh-Guldberg O (2011) Regulation of apoptotic mediators reveals dynamic responses to thermal stress in the reef building coral *Acropora millepora*. PLoS One 6:e16095

Pochon X, Gates RD (2010) A new *Symbiodinium* clade (Dinophyceae) from soritid foraminifera in Hawai'i. Mol Phylogenet Evol 56:492–497

Pochon X, Stat M, Takabayashi M, Chasqui L, Chauka LJ, Logan DDK, Gates RD (2010) Comparison of endosymbiotic and free-living *Symbiodinium* (Dinophyceae) diversity in a Hawaiian reef environment. J Phycol 46:53–65

Pochon X, Putnam HM, Burki F, Gates RD (2012) Identifying and characterizing alternative molecular markers for the symbiotic and free-living dinoflagellate genus *Symbiodinium*. PLoS One 7(1): e29816

Potts DC, Done TJ, Isdale PJ, Fisk DA (1985) Dominance of a coral community by the genus *Porites* (Scleractinia). Mar Ecol Prog Ser 23:79–84

Putnam HM, Stat M, Pochon X, Gates RD (2012) Endosymbiotic flexibility associates with environmental sensitivity in scleractinian corals. Proc Roy Soc B 279:4352–4361

Ralph PJ, Gademann R, Larkum AWD (2001) Zooxanthellae expelled from bleached corals at 33°C are photosynthetically competent. Mar Ecol Prog Ser 220:163–168

Ralph PJ, Larkum AWD, Kuhl M (2005) Temporal patterns in effective quantum yield of individual zooxanthellae expelled during bleaching. J Exp Mar Biol Ecol 316:17–28

Richier S, Sabourault C, Courtiade J, Zucchini N, Allemand D, Furla P (2006) Oxidative stress and apoptotic events during thermal stress in the symbiotic sea anemone, *Anemonia viridis*. FEBS J 273:4186–4198

Rodriguez-Lanetty M, Harii S, Hoegh-Guldberg O (2009) Early molecular responses of coral larvae to hyperthermal stress. Mol Ecol 18:5101–5114

Rosic NN (2012) Phylogenetic analysis of genes involved in mycosporine-like amino acid biosynthesis in symbiotic dinoflagellates. Appl Microbiol Biotechnol 94:29–37

Roth MS (2014) The engine of the reef: photobiology of the coral-algal symbiosis. Front Microbiol 5:422

Rowan R (2004) Coral bleaching: thermal adaptation in reef coral symbionts. Nature 430:742

Rowan R, Powers DA (1991a) A molecular genetic classification of zooxanthellae and the evolution of animal-algal symbiosis. Science 251:1348–1351

Rowan R, Powers DA (1991b) Molecular genetic identification of symbiotic dinoflagellates (zooxanthellae). Mar Ecol Prog Ser 71:65–73

Salih A, Larkum A, Cox G, Kuhl M, Hoegh-Guldberg O (2000) Fluorescent pigments in corals are photoprotective. Nature 408:850–853

Sammarco PW (1982) Polyp bail-out: an escape response to environmental stress and a new means of reproduction in corals. Mar Ecol Prog Ser 10:57–65

Sampayo EM, Ridgway T, Bongaerts P, Hoegh-Guldberg O (2008) Bleaching susceptibility and mortality of corals are determined by fine-scale differences in symbiont type. Proc Natl Acad Sci U S A 105:10444–10449

Sampayo EM, Dove S, LaJeunesse TC (2009) Cohesive molecular genetic data delineate species diversity in the dinoflagellate genus *Symbiodinium*. Mol Ecol 18:500–519

Santos SR, Coffroth MA (2003) Molecular genetic evidence that dinoflagellates belonging to the genus *Symbiodinium* Freudenthal are haploid. Biol Bull 204:10–20

Saragosti E, Tchernov D, Katsir A, Shaked Y (2010) Extracellular production and degradation of superoxide in the coral *Stylophora pistillata* and cultured *Symbiodinium*. PLoS One 5:e12508

Schwarz JA, Krupp DA, Weis VM (1999) Late larval development and onset of symbiosis in the scleractinian coral *Fungia scutaria*. Biol Bull 196:70–79

Shikina S, Chen CJ, Liou JY, Shao ZF, Chung YJ, Lee YH, Chang CF (2012) Germ cell development in the scleractinian coral *Euphyllia ancora* (Cnidaria, Anthozoa). PLoS One 7(7):e41569

Shinzato C, Shoguchi E, Kawashima T, Hamada M, Hisata K, Tanaka M, Fujie M, Fujiwara M, Koyanagi R, Ikuta T, Fujiyama A, Miller DJ, Satoh N (2011) Using the *Acropora digitifera* genome to understand coral responses to environmental change. Nature 476:320–324

Shinzato C, Shoguchi E, Tanaka M, Satoh N (2012) Fluorescent protein candidate genes in the coral *Acropora digitifera* genome. Zool Sci 29:260–264

Silverstein RN, Correa AMS, Baker AC (2012) Specificity is rarely absolute in coral-algal symbiosis: implications for coral response to climate change. Proc R Soc B 279:2609–2618

Smith LD, Hughes TP (1999) An experimental assessment of survival, re-attachment and fecundity of coral fragments. J Exp Mar Biol Ecol 235:147–164

Smith-Keune C, Dove S (2008) Gene expression of a green fluorescent protein homolog as a host-specific biomarker of heat stress within a reef-building coral. Mar Biotechnol 10:166–180

Stanley GD Jr (2006) Photosymbiosis and the evolution of modern coral reefs. Science 312:857–858

Stanley GD, Swart PK (1995) Evolution of the coral-zooxanthellae symbiosis during the Triassic: a geochemical approach. Paleobiology 21:179–199

Stat M, Loh WKW, LaJeunesse TC, Hoegh-Guldberg O, Carter DA (2009) Stability of coral-endosymbiont associations during and after a thermal stress event in the southern Great Barrier Reef. Coral Reefs 28:709–713

Stat M, Bird CE, Pochon X, Chasqui L, Chauka LJ, Concepcion GT, Loga D, Takabayashi M, Toonen RJ, Gates RD (2011) Variation in *Symbiodinium* ITS2 sequence assemblages among coral colonies. PLoS One 6(1):e15854

Stoddart JA (1983) Asexual production of planulae in the coral *Pocillopora damicornis*. Mar Biol 76:279–284

Strychar KB, Sammarco PW (2008) Exaptation in corals to high seawater temperatures: low concentrations of apoptotic and necrotic cells in host coral tissue under bleaching conditions. J Exp Mar Biol Ecol 369:31–42

Strychar KB, Coates M, Sammarco PW, Piva TJ (2004) Bleaching as a pathogenic response in scleractinian corals, evidenced by high concentrations of apoptotic and necrotic zooxanthellae. J Exp Mar Biol Ecol 304:99–121

Sunagawa S, Wilson EC, Thaler M, Smith ML, Ccaruso C, Pringle JR, Weis VM, Medina M, Schwarz JA (2009) Generation and analysis of transcriptomic resources for a model system on the rise: the sea anemone *Aiptasia pallida* and its dinoflagellate endosymbiont. BMC Genomics 10:258

Suwa R, Hirose M, Hidaka M (2008) Seasonal fluctuation in zooxanthella composition and photo-physiology in the corals *Pavona divaricata* and *P. decussata* in Okinawa. Mar Ecol Prog Ser 361:129–137

Szmant-Froelich AM, Reutter M, Riggs L (1985) Sexual reproduction of *Favia fragum* (Esper): lunar patterns of gametogenesis, embryogenesis and planulation in Puerto Rico. Bull Mar Sci 37:880–892

Taguchi T, Mezaki T, Iwase F, Sekida S, Kubota S, Fukami H, Okuda K, Shinbo T, Oshima S, Iiguni Y, Testa JR, Tominaga A (2014) Molecular cytogenetic analysis of the scleractinian coral *Acropora solitaryensis* Veron & Wallace 1984. Zool Sci 31:89–94

Takahashi S, Nakamura T, Sakamizu M, van Woesik R, Yamasaki H (2004) Repair machinery of symbiotic photosynthesis as the primary target of heat stress for reef-building corals. Plant Cell Physiol 45:251–255

Takahashi S, Whitney SM, Badger MR (2009) Different thermal sensitivity of the repair of photodamaged photosynthetic machinery in cultured *Symbiodinium* species. Proc Natl Acad Sci U S A 106:3237–3242

Tchernov D, Gorbunov MY, de Vargas C, Yadav SN, Milligan AJ, Haggblom M, Falkowski PG (2004) Membrane lipids of symbiotic algae are diagnostic of sensitivity to thermal bleaching in corals. Proc Natl Acad Sci U S A 101:13531–13535

Tchernov D, Kvitt H, Haramaty L, Bibby TS, Gorbunov MY, Rosenfeld H, Falkowsky PG (2011) Apoptosis and the selective survival of host animals following thermal bleaching in zooxanthellate corals. Proc Natl Acad Sci U S A 108:9905–9909

Thornhill DJ, LaJeunesse TC, Kemp DW, Fitt WK, Schmidt GW (2005) Multi-year, seasonal genotypic surveys of coral-algal symbioses reveal prevalent stability or post-bleaching reversion. Mar Biol. doi:10.1007/s00227-005-0114-2

Trench RK, Blank RJ (1987) *Symbiodinium microadriaticum* Freudenthal, *S. goreauii* sp. nov., *S. kawagutii* sp. nov. and *S. pilosum* sp. nov.: gymnodinioid dinoflagellate symbionts of marine invertebrates. J Phycol 23:469–481

Tsuta H, Hidaka M (2013) Telomere length of the colonial coral *Galaxea fascicularis* at different developmental stages. Coral Reefs 32:495–502

Tsuta H, Shinzato C, Satoh N, Michio Hidaka M (2014) Telomere shortening in the colonial coral *Acropora digitifera* during Development. Zool Sci 31:129–134

Twan W-H, Hwang J-S, Lee Y-H, Wu H-F, Tung Y-H, Chang C-F (2006) Hormones and reproduction in scleractinian corals. Comp Biochem Physiol A 144:247–253

Udy JW, Hinde R, Vesk M (1993) Chromosomes and DNA in *Symbiodinium* from Australian hosts. J Phycol 29:314–320

Van Oppen MJH, Gates RD (2007) Conservation genetics and the resilience of reef-building corals. Mol Ecol 15:3863–3883

Van Oppen MJH, Souter P, Howells EJ, Heyward A, Berkelmans R (2011) Novel genetic diversity through somatic mutations: fuel for adaptation of reef corals? Diversity 3:405–423

Vizel M, Loya Y, Downs C, Kramarsky-Winter E (2011) A novel method for coral explant culture and micropropagation. Mar Biotechnol 13:423–432

Voolstra CR, Schwarz JA, Schnetzer J, Sunagawa S, Desalvo MK, Szmant AM, Coffroth MA, Medina M (2009) The host transcriptome remains unaltered during the establishment of coral-algal symbioses. Mol Ecol 18:1823–1833

Warner ME, Fitt WK, Schmidt G (1999) Damage to photosystem II in symbiotic dinoflagellates: a determinant of coral bleaching. Proc Natl Acad Sci U S A 96:8007–8012

Watanabe H, Hoang VT, Mättner R, Holstein TW (2009) Immortality and the base of multicellular life: lessons from cnidarians stem cells. Semin Cell Dev Biol 20:1114–1125

Weis VM (2008) Cellular mechanisms of Cnidarian bleaching: stress causes the collapse of symbiosis. J Exp Biol 211:3059–3066

Wewengkang DS, Watanabe T, Hidaka M (2007) Studies on morphotypes of the coral *Galaxea fascicularis* from Okinawa: polyp color, nematocyst shape, and coenosteum density. Galaxea J Coral Reef Stud 9:49–59

Yakovleva I, Hidaka M (2004) Differential recovery of PSII function and electron transport rate in symbiotic dinoflagellates as possible determinant of bleaching susceptibility of corals. Mar Ecol Prog Ser 268:43–53

Yakovleva IM, Bhagooli R, Takemura A, Hidaka M (2004) Differential susceptibility to oxidative stress of two scleractinian corals: antioxidant functioning of mycosporine-glycine. Comp Biochem Physiol B 139:721–730

Yakovleva IM, Baird AH, Yamamoto HH, Bhagooli R, Nonaka M, Hidaka M (2009) Algal symbionts increase oxidative damage and death in coral larvae at high temperature. Mar Ecol Prog Ser 378:105–112

Yamashiro H, Hidaka M, Nishihira M, Poung-In S (1989) Morphological studies on skeletons of *Diaseris fragilis*, a free-living coral which reproduces asexually by natural autotomy. Galaxea 8:283–294

Yamashita H, Koike K (2013) Genetic identity of free-living *Symbiodinium* obtained over a broad latitudinal range in the Japanese coast. Phycol Res 61:68–80

Yeoh SR, Dai CF (2010) The production of sexual and asexual larvae within single broods of the scleractinian coral, *Pocillopora damicornis*. Mar Biol 157:351–359

Yuyama I, Watanabe T, Takei Y (2010) Profiling differential gene expression of symbiotic and aposymbiotic corals using a high coverage expression profiling (HiCEP) analysis. Mar Biotechnol. doi:10.1007/s10126-010-9265-3

Yuyama I, Ito Y, Watanabe T, Hidak M, Suzuki Y, Nishida M (2012) Differential gene expression in juvenile polyps of the coral *Acropora tenuis* exposed to thermal and chemical stresses. J Exp Mar Biol Ecol 430–431:17–24

Chemical and Biological Characteristics of Coral Reef Ecosystem at Microscale/Nanoscale: Effect of Multiple and Synergistic Stresses

2

Beatriz E. Casareto, Toshiyuki Suzuki, and Yoshimi Suzuki

Abstract

Global environmental changes are recently occurring faster than any other time. The resulting combinations of natural and anthropogenic disturbances are strongly affecting coral reef communities. Coral damage can be caused by both abiotic factors (temperature, sedimentation, nutrients inputs, ultraviolet radiation) and biotic factors (predation, overgrowth of algae, infectious diseases). These factors acting mostly in synergy had resulted in worldwide coral reef deterioration. Coral bleaching is the most impacting process that is affecting coral survival under elevated sea surface temperature and high irradiance scenario. Bleaching is well known to occur around the world; however, its mechanism is not well understood. This is due to the high complexity of the "coral holobiont" (coral in symbiosis with its zooxanthellae and a microbial community that maintains a delicate balance to keep the coral health). In this chapter a novel point of view of bleaching mechanism using micro/nano-size scales is presented: (1) the study of pigment dynamics during thermal-induced bleaching revealed that bleaching is a detoxification strategy to avoid the formation of reactive oxygen species (ROS), (2) the synergistic action of thermal stress with pathogenic bacteria exacerbates the bleaching process, and (3) the synergistic effect of thermal stress in a nitrate-enriched environment can impede the recovery of corals after a bleaching event, turning the corals to be more susceptible to other environmental or anthropogenic stressors.

Keywords

Coral bleaching • Bacteria • Nutrient • Pigment

2.1 New Approach for Understanding Coral Bleaching Mechanism Using Pigment Analysis

2.1.1 Introduction

Previously it was thought that bleaching was the result of expulsion of zooxanthellae by the host (Hoegh-Guldberg and Smith 1989; Gates 1990; Brown et al. 1995; Jones 1997), or it resulted from the degradation of photosynthetic pigments in zooxanthellar cells (Fitt and Warner 1995; Fitt et al. 2001). The loss of zooxanthellae has been linked to several environmental stressors, especially susceptibility to

The original version of this chapter was revised. An erratum to this chapter can be found at DOI 10.1007/978-4-431-54364-0_9.

B.E. Casareto (✉)
Research Institute of Green Science and Technology, Shizuoka University, 836 Ohya, Suruga-ku, Shizuoka 422-8529, Shizuoka, Japan

Graduate School of Science and Technology, Shizuoka University, 836 Ohya, Suruga-ku, Shizuoka 422-8529, Shizuoka, Japan
e-mail: becasar@shizuoka.ac.jp

T. Suzuki • Y. Suzuki
Graduate School of Science and Technology, Shizuoka University, 836 Ohya, Suruga-ku, Shizuoka 422-8529, Shizuoka, Japan
e-mail: suzuki.toshiyuki@shizuoka.ac.jp; suzuki.yoshimi@shizuoka.ac.jp

H. Kayanne (ed.), *Coral Reef Science*, Coral Reefs of the World 5, DOI 10.1007/978-4-431-54364-0_2

high water temperature, which damages cells by changing the chloroplast morphology and function (Bhagooli and Hidaka 2002, 2003, 2006). Kuroki and van Woesik (1999) reported that during the massive Okinawan bleaching event in 1998, zooxanthellae retained in coral tissues changed morphologically and lost pigmentation. Also other authors (Mise and Hidaka 2003; Reimer et al. 2007) described diverse forms of zooxanthellae in tissues of naturally bleached coral during summer.

To examine the morphology and abundance of expelled and retained zooxanthellae, temperature stresses were applied to the coral *Montipora digitata*. Expelled algal cells were collected for classification, enumeration, and analysis of their pigments using high-performance liquid chromatography (HPLC). These parameters were used to compare retained and expelled cells, paying particular attention to pigment composition shifts in response to thermal stress.

2.1.2 Degradation of Zooxanthellae in Coral

Branches of *M. digitata* were collected from a single colony growing in the waters of Bise, Motobu, Okinawa, Japan (26°42′N and 127°52′E). Three branches were incubated in glass bottles containing 800 ml of filtered (0.2 μm) seawater at 12-h light and 12-h dark illumination condition. Incubations were done at 27 °C (control) and 32 °C. To make observations on expelled zooxanthellae, water from incubation vessel was collected during daytime and nighttime. Half of the collected water was filtered through 2.0-μm Nuclepore polycarbonate membranes for observation and counting of expelled zooxanthellae; the remaining water was passed through GF/F filters for pigment analyses. Zooxanthellae in tissue were collected by removing coral tissue using Waterpik.

We classified zooxanthellae into three categories: (i) healthy morphology with normally expanded chloroplasts; (ii) shrunken cells of reduced size with partially fragmented, darkened, and shrunken chloroplasts; and (iii) bleached cells with pale or colorless chloroplasts (Fig. 2.1). We observed three zooxanthellar morphologies in coral tissue, but bleached cells were rare (0.39 % of totals at 27 °C and 1.97 % at 32 °C). As indicated in Table 2.1, the density of zooxanthellae in coral branches after 4 days of incubation at 27 °C was similar to the initial value, as were the proportions of shrunken and healthy zooxanthellae. At 32 °C, zooxanthellar density declined significantly (t-test, $p = 0.002$) to 42 % of the initial value, and the number of shrunken cells increased from 3.78×10^4 to 4.25×10^5 cells cm^{-2}, accounting for ~18 % of total density. Zooxanthellar expulsion rates at the two temperatures are detailed in Table 2.2 and Fig. 2.2. At 27 °C, expelled cell numbers ranged from 3.78×10^2 to 2.39×10^3 cells cm^{-2} coral during 12 h of darkness and from 3.06×10^3 to 1.82×10^4 cells cm^{-2} coral during 12 h of illumination (Fig. 2.2). More zooxanthellae were expelled during illumination and most were shrunken. At 32 °C, expelled cell numbers ranged from 2.27×10^2 to 1.41×10^3 cells cm^{-2} coral during 12 h of darkness and from 5.47×10^2 to 9.87×10^2 cells cm^{-2} coral during 12 h of illumination. Total numbers of expelled cells over 4 days were 4.39×10^4 at 27 °C and 6.00×10^3 at 32 °C, accounting for ~1 % of the total zooxanthellae contained in coral tissue at the outset.

Since the number of cells expelled was very low in the incubations, it was proposed that the expulsion of zooxanthellae from coral is a natural physiological phenomenon and may not be the main mechanism underlying coral bleaching. Although zooxanthellae were expelled at a low rate from coral held at 32 °C, the algal cell density greatly decreased inside coral tissues. A decrease in the zooxanthellar mitotic index in corals held at 32 °C (compared to 27 °C) was also observed. Therefore, algal cell division was reduced by thermal stress, but this does not explain the decrease in cell density inside the coral tissue (Table 2.1): there must be a process of zooxanthellar degradation inside the host (Fig. 2.3). Titlyanov et al. (1996) reported that coral hosts commonly digest their algal symbionts, a phenomenon that has also been observed in the sea anemone *Phyllactis flosculi* (Steele and Goreau 1977), in giant clams (Fankboner 1971), and in the marine hydroid *Myrionema amboinense* (Fitt and Cook 1990). In this research, a large number of shrunken zooxanthellae in coral tissue and in the water of the experimental system were found. Zooxanthellae with shrunken cytoplasm and reduced chloroplasts have been observed previously in corals under thermal stress (Fukabori 1998) and several coral species, e.g., *M. digitata* (Titlyanov et al. 1996; Papina et al. 2007), *Stylophora pistillata* (Titlyanov et al. 1996; Kuroki and van Woesik 1999; Titlyanov et al. 2001), *Galaxea fascicularis* (Bhagooli and Hidaka 2002), and *Zoanthus sansibaricus* (Reimer et al. 2007), among others (*Acropora selago*, *Acropora muricata*, *Heliofungia actiniformis*, *Ctenactis echinata*, *Oxypora lacera*, and *Pocillopora eydouxi*; Fujise et al. 2013). Although these zooxanthellae have been classified as degraded (Titlyanov et al. 1998; Downs et al. 2009, 2013), the mechanism responsible for the formation of these shrunken cells is poorly understood.

2.1.3 Finding of Cyclo Enol in Shrunken Zooxanthellae

Shrunken zooxanthellae collected from the seawater in the incubation vessel at 27 °C were compared in their pigments with those of healthy algal cells collected from coral tissue.

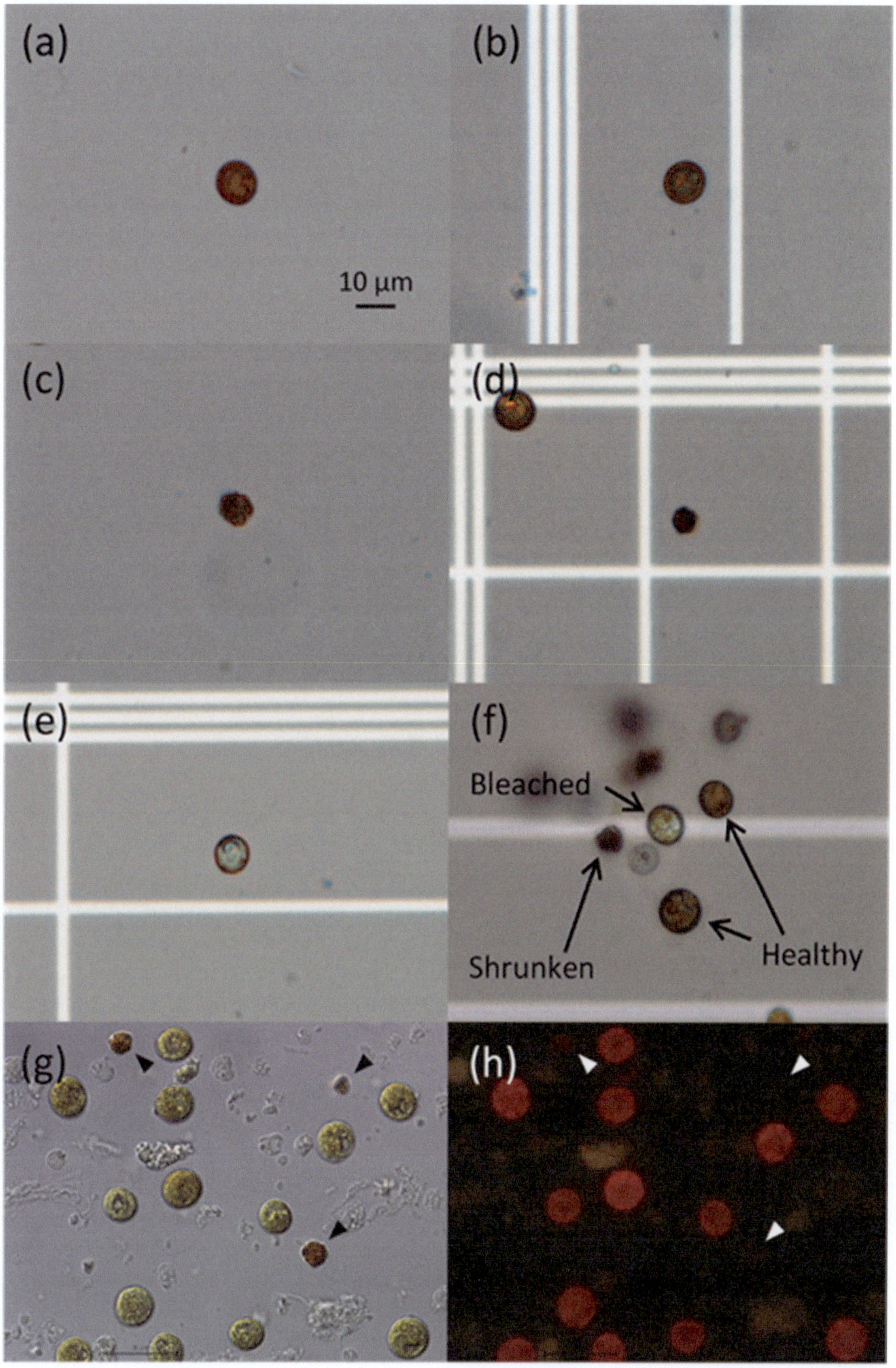

Fig. 2.1 Types of zooxanthellae observed in coral tissue. (**a**, **b**) Healthy cells with a spherical shape and expanded chloroplast. (**c**, **d**) Shrunken, darkly colored cells with reduced sizes and partially fragmented chloroplasts. (**e**) Bleached cells with pale and colorless chloroplasts. (**f**) Three categories of zooxanthellae. (**g**) White light micrograph and (**h**) fluorescence image of healthy and shrunken cells (shrunken cells are indicated by arrowheads) (Reproduced from Suzuki et al. (2015) by permission of John Wiley & Sons Ltd. Copyright © 2014 The Authors. *Journal of Phycology* published by Wiley Periodicals, Inc. on behalf of Phycological Society of America)

Elution profiles of pigments of shrunken and healthy zooxanthellae are depicted in Fig. 2.4. Seven peaks appeared only in samples dominated by shrunken zooxanthellae: pheophorbide *a* and pigments similar to it (17.05-min retention time), a pigment similar to peridinin (22.33 and 22.69 min), a pigment similar to diadinochrome

Table 2.1 Numbers and proportions of zooxanthellae (cells cm^{-2} coral surface) retained and expelled from coral tissue after 4 days of incubation at two temperatures

Zooxanthellae retained			
	Day 0	Day 4	
		(27 °C)	(32 °C)
Total	5.64×10^6	5.65×10^6	2.37×10^6
Healthy	5.60×10^6 (99.3 %)	5.52×10^6 (97.7 %)	1.95×10^6 (82.3 %)
Shrunken	3.78×10^4 (0.7 %)	1.25×10^5 (2.3 %)	4.25×10^5 (17.7 %)
Zooxanthellae expelled over 4 days			
		(27 °C)	(32 °C)
Total		4.39×10^4	6.00×10^3
Healthy		8.55×10^3 (19.4 %)	3.08×10^3 (51.3 %)
Shrunken		3.54×10^4 (80.6 %)	2.92×10^3 (48.7 %)

Table 2.2 Rates of zooxanthellar expulsion (cells cm^{-2} coral surface h^{-1}) at two temperatures during illuminated and dark periods of the day

	Total	Healthy	Shrunken
(27 °C)			
Day	808.2	124.7	683.5
Night	106.5	53.4	53.1
(32 °C)			
Day	61.5	30.6	30.9
Night	63.5	33.6	29.8

(27.76 min), a type of chlorophyll (31.03 min), a pigment similar to alloxanthin (32.13 min), and pyropheophytin *a* (39.93 min). A noticeable pigment peak at 31.03-min retention time had a maximum absorption peak at 686 nm (red band), matching a report by Goericke et al. (2000). The absorption spectrum of this pigment (extracted from zooxanthellae) was compared with an authentic standard of 13^2, 17^3-cyclopheophorbide *a* enol (cyclo enol, cPPB-*a*E) and almost perfect match, and therefore the extracted pigment was identified as cyclo enol (on the basis of retention time and absorption spectrum). Concentrations of chlorophyll *a*, peridinin, and chlorophyll c_2 after 4 days were closely similar at the two temperatures (Fig. 2.5); however, the concentration of cyclo enol was much higher at 32 °C (Fig. 2.5), even though cell numbers had declined at this higher temperature (Table 2.1). Pigment contents of expelled zooxanthellae are shown in Fig. 2.6. Chlorophyll *a* and c_2 concentrations were low at both temperatures. Cyclo enol was the most abundant pigment extracted from zooxanthellae expelled from corals at 27 °C (Fig. 2.6), and the number of expelled shrunken zooxanthellae exceeded the number of healthy cells (Fig. 2.4).

Cyclo enol has been reported as a degradation product of chlorophyll *a* in phytoplankton (Kashiyama et al. 2012), and it is commonly present in aquatic environments: marine and lacustrine sediments (Chillier et al. 1993; Harris et al. 1995; Ocampo et al. 1999; Louda et al. 2000), sponges (Karuso et al. 1986), bivalves (Sakata et al. 1990; Yamamoto et al. 1992; Watanabe et al. 1993; Louda et al. 2008), and protozoa (Goericke et al. 2000). Kashiyama et al. (2012, 2013) found that herbivorous protozoa produce cyclo enol when they graze on and digest microalgae. Cyclo enol is generated from pyropheophytin *a* (Kashiyama and Tamiaki 2014). Several types of phytoplankton are also able to generate cyclo enol (Kashiyama et al. 2014). Yamada et al. (2013) reported the formation of cyclo enol (in small quantities) in zooxanthellae that had been extracted from coral and cultivated in flask in stationary phase. In this research, cyclo enol and pyropheophytin *a* were detected in pigment extracts of shrunken zooxanthellae. Thus, cyclo enol is likely generated from chlorophyll *a* through a degradation pathway leading to the formation of shrunken zooxanthellae.

Degradation of zooxanthellae inside the coral tissues is still a question. It is hypothesized that this process may be a detoxification strategy to avoid reactive oxygen species (ROS) formation. Free chlorophyll *a* released from broken chloroplasts becomes a generator of singlet oxygen when exposed to light, thereby promoting the formation of ROS which results in significant damage of cell structure (Perl-Treves and Perl 2002). Cyclo enol differs from chlorophyll *a* in that it does not have fluorescence; therefore, ROS are not formed (Kashiyama et al. 2012, 2013). Protozoans that feed on microalgae have transparent bodies, and their body contents are therefore always exposed to light during daytime. Accordingly, these organisms have developed a strategy for detoxifying free chlorophyll *a* through degradation to the non-fluorescing product cyclo enol (Kashiyama et al. 2012, 2013). In case of coral, the red fluorescence of chlorophyll was largely quenched in the shrunken algal cells (Fig. 2.1g, h). Kashiyama et al. (2012) also reported that chloroplasts of diatoms grazed on by protozoans were shrunken and had no chlorophyll fluorescence. This loss of fluorescence may indicate that ROS are not produced by chlorophylls freed from damaged chloroplasts. Corals also have transparent bodies and they live symbiotically with zooxanthellae. Therefore, they are always exposed to potential damage caused by the oxidative stress of ROS (Lesser et al. 1990; Dykens et al. 1992; Downs et al. 2002). Oxidative damage becomes more severe as UV radiation and water temperature increase (Lesser et al. 1990). Moreover, damaged chloroplasts are repaired with difficulty during thermal stress episodes, and ROS formation is thereby increased (Bhagooli and Hidaka 2006). It is proposed that corals and zooxanthellae employ the detoxification strategy used by herbivorous protists and phytoplankton, viz., degradation of chlorophyll *a* to cyclo enol. Furthermore, reductions in zooxanthellar numbers within coral tissue caused by

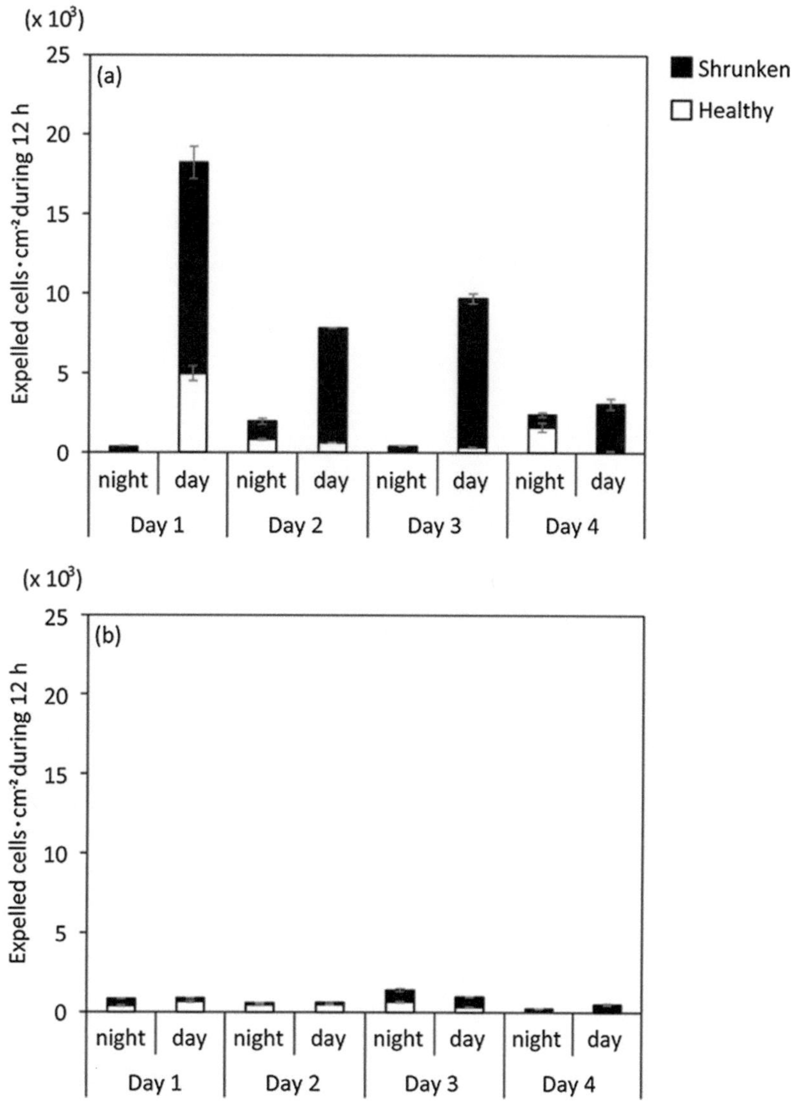

Fig. 2.2 Zooxanthellar density and composition in water from incubation vessel collected over 12-h intervals for 4 days. (**a**) 27 °C and (**b**) 32 °C. Cell numbers are normalized to surface areas of coral branches. Values are means±SE ($n = 3$) (Reproduced from Suzuki et al. (2015) by permission of John Wiley & Sons Ltd. Copyright © 2014 The Authors. *Journal of Phycology* published by Wiley Periodicals, Inc. on behalf of Phycological Society of America)

degradation of algal cells and bleaching may be an important mechanism for reducing the production of ROS (Fig. 2.7). Thus, coral bleaching is a physiological mechanism used as a survival strategy by corals facing oxidative damage.

2.1.4 Summary of Sect. 2.1

The morphology and pigment composition of zooxanthellae were examined in corals subjected to normal temperature (27 °C) and thermal stress (32 °C). Several normal and abnormal morphological types of zooxanthellar cells were observed. Normal cells were intact and their chloroplasts were unbroken (healthy); abnormal cells were shrunken and had partially degraded or broken chloroplasts, or they were bleached without visible chloroplasts. At 27 °C, most healthy zooxanthellae were retained in the coral tissue, whereas shrunken zooxanthellae were expelled. Under thermal stress, the abundance of healthy zooxanthellae declined and the proportion of shrunken/abnormal cells increased in coral tissues. Within the shrunken cells, it was detected the presence of a chlorophyll-like pigment that is not ordinarily found in healthy zooxanthellae. Analysis of the absorption spectrum, absorption maxima, and retention time indicated that this pigment was 13^2, 17^3-cyclopheophorbide *a* enol (cyclo enol, cPPB-*a*E), which is frequently found in marine and lacustrine sediments and in protozoans that graze on phytoplankton. The production of cyclo enol in shrunken

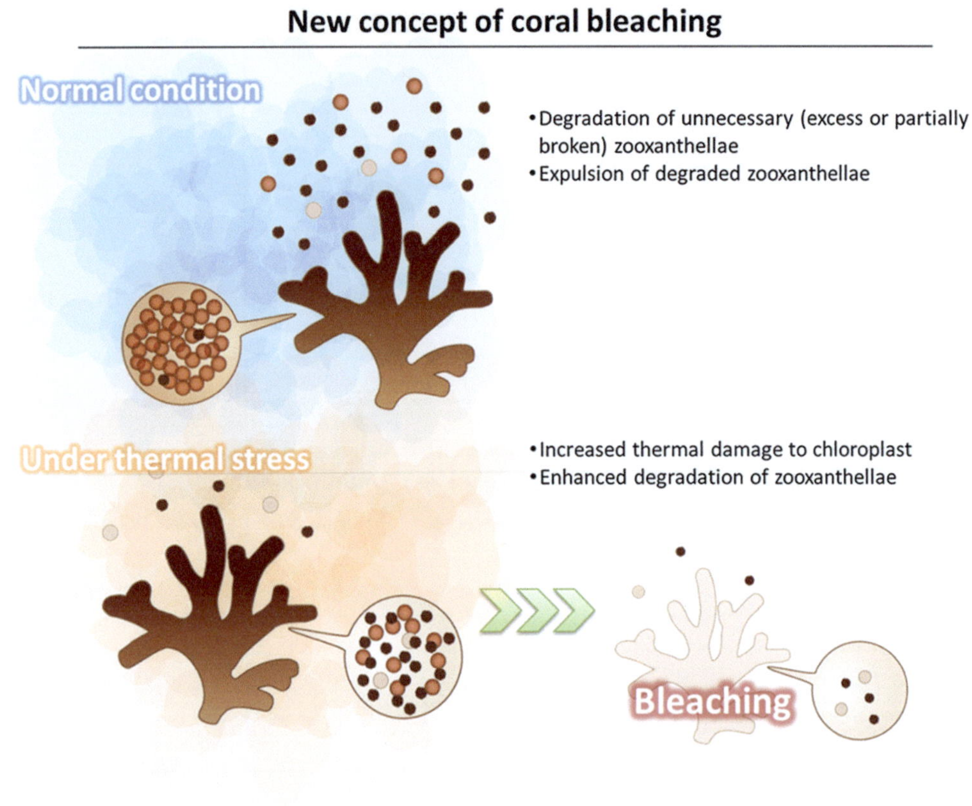

Fig. 2.3 New concept of coral bleaching revealed from counting and observation of zooxanthellae

zooxanthellae suggests that the chlorophylls have been degraded to cyclo enol, a compound that is not fluorescent. The lack of a fluorescence function precludes the formation of reactive oxygen species. Therefore, it was considered that the formation of cyclo enol in shrunken zooxanthellae is a mechanism for avoiding oxidative stress.

2.2 Enhancement of Bleaching Under Synergistic Action of Thermal Stress and Pathogenic Bacteria on the Coral *Montipora digitata*

2.2.1 Introduction

Corals are complex symbiotic systems including the host, a cnidarian, the endosymbiotic algae, prokaryotes (bacteria and archaea), fungi, and virus. Although the role of the microorganisms associated with corals is not yet well defined, they seem to play an important role in the maintenance of the coral health by providing essential nutrients (Agostini et al. 2009) and defense against potential pathogens (Ritchie 2006). Their potential roles in the maintenance of coral health were reviewed in Rosenberg et al. (2007). Beneficial roles of the associated prokaryotes have been reported in the literature, but on the other side, growing evidences show the existence of coral pathogens which can cause severe diseases often resulting in the death of the coral. Bleaching of the coral *Oculina patagonica* in the Mediterranean Sea has been shown to be caused by the coral pathogen *Vibrio shiloi* (Kushmaro et al. 1997). In *Pocillopora damicornis*, the coral pathogen *Vibrio coralliilyticus* causes bleaching followed by tissue necrosis (Ben-Haim et al. 2003). Moreover, there is a higher prevalence of disease following bleaching events (Muller et al. 2007). These results may be due to a shift in the coral-associated microbial community during a bleaching event (Bourne et al. 2008), changes in the coral metabolism under high-temperature stress especially the decrease in primary production (Fujimura et al. 2008), and the release of mucus and ammonia (Suzuki and Casareto 2011). Therefore, the effects of high seawater temperature and potential pathogens on the coral metabolism should be studied to understand the synergistic effects of these two stresses and the sequences that lead from a temperature-induced bleaching event to bacteria/temperature-caused bleaching event along with their differences. Coral bleaching triggered by a thermal stress disrupts the function of the coral metabolism, weakening the coral and allowing opportunistic pathogens to cause damages to the host. Figure 2.8 illustrates

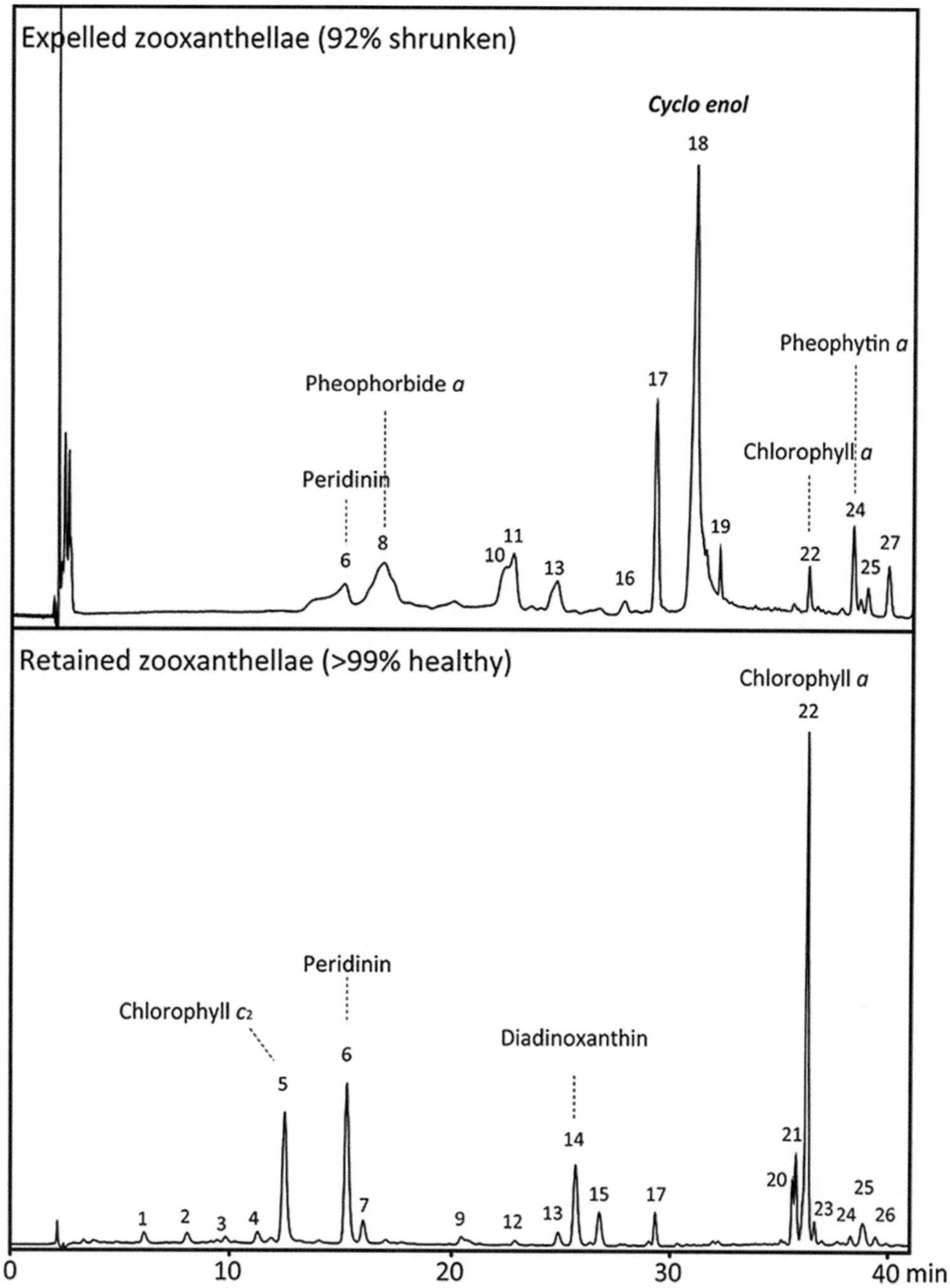

Fig. 2.4 Elution profiles of expelled zooxanthellae (*upper*) and retained zooxanthellae at the outset of the experiment (*lower*) at 27 °C (Reproduced from Suzuki et al. (2015) by permission of John Wiley & Sons Ltd. Copyright © 2014 The Authors. *Journal of Phycology* published by Wiley Periodicals, Inc. on behalf of Phycological Society of America

three possible ways of coral bleaching considering different stages of coral health from a normal stage, passing through a step where bleaching has already started affecting some of the metabolic functions of corals and/or the associated zooxanthellae but still without any visible signals, followed by a step of visible bleaching and a serious compromised tissue necrosis that takes the coral finally to death. All these coral health stages can be imagined through an arbitrary timescale depending on the bleaching type as follows:

(a) Thermal stress-mediated bleaching with a possible recovering if the thermal stress ceases, with a relatively long timescale.
(b) Microbial mediated bleaching (apply for the case studied on the coral *Pocillopora damicornis* triggered by the bacteria *V. coralliilyticus*). This process is almost irreversible and takes the coral to death in rapid timescale.
(c) Synergistic action of thermal stress and pathogenic bacteria infection: this process results from the infection of

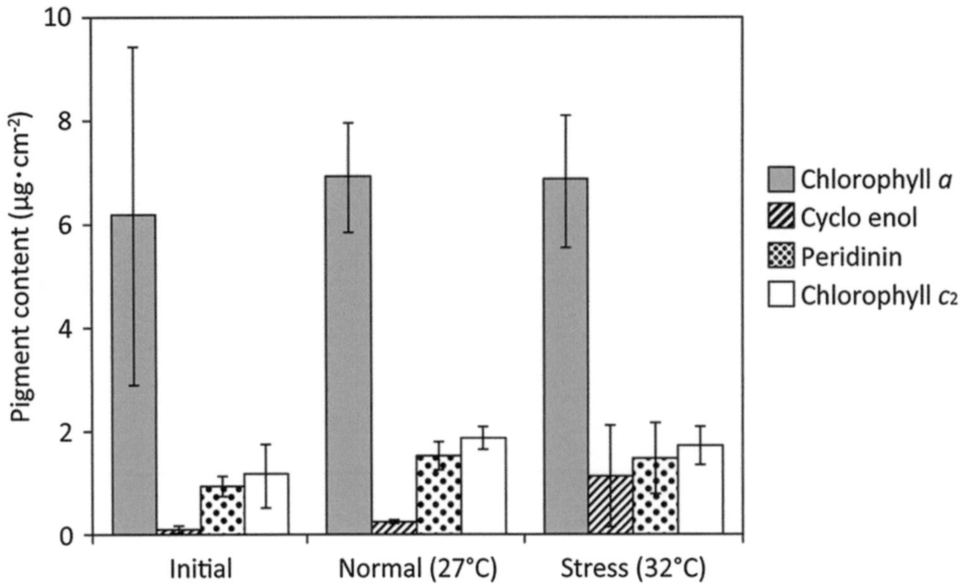

Fig. 2.5 Initial pigment concentrations of zooxanthellae retained in coral tissue compared with values after 4 days of incubation at 27 °C and 32 °C. Pigment contents are normalized to surface areas of coral branches. Values are means±SD ($n = 9$) (Reproduced from Suzuki et al. (2015) by permission of John Wiley & Sons Ltd. Copyright © 2014 The Authors. *Journal of Phycology* published by Wiley Periodicals, Inc. on behalf of Phycological Society of America)

coral during bleaching induced by thermal stress. This can be explained due to the weakness of coral once some changes in metabolic functions start to occur due the thermal stress, facilitating infection of opportunistic pathogens.

The present research tries to describe the (c) type of synergistic actions of thermal stress combined with pathogen bacteria.

2.2.2 Approach for Testing Bacteria Strains in an Incubation

Mixed solutions containing five species of bacteria, *Vibrio coralliilyticus* (AB490821), *Vibrio harveyi* (AB490822), *Paracoccus carotinifaciens* (AB490820), *Pseudoalteromonas* sp. (AB691769), and *Sulfitobacter* sp. (AB691770), were prepared at the concentration of 5.5×10^5 ml^{-1} to 1.2×10^6 ml^{-1} and inoculated to coral branches of the coral *Montipora digitata* kept in incubation vessels (four branches/vessel in triplicates). To confirm the effects of each bacterial species, individual species were also tested, using a final abundance of approximately 10^6 cells ml^{-1}. These bacteria are naturally found in coastal regions of Okinawa, Japan.

Coral branches of the coral *Montipora digitata* were kept in incubation vessels (four branches/vessel in triplicates). Seawater temperatures were set at 27 °C or 32 °C, and two levels of bacterial abundance (no addition and addition of mixed bacteria) were tested in a 2 × 2 factorial design with three replicates per treatment. Continuous flow of seawater was provided to incubation vessels at the rate of 15 ml min^{-1}. Illumination was provided at 12:12-h light/dark cycle. A total of 30 glass bottles were prepared: 15 bottles were kept in a water bath maintained at a constant temperature of 27 °C, and the other 15 bottles were in a water bath at 32 °C. After 4 days, the bacteria were inoculated into six bottles at each temperature, using an additional pump to maintain a total flow rate of 15 ml min^{-1}. Tissue samples and water samples for measurements of coral metabolisms were taken on day 0 (initial), day 4, and day 8. Corals from three replicate bottles under each condition were used for measurements of photosynthetic efficiency (Fv/Fm) and posteriorly sacrificed for tissue analysis (zooxanthellae counts, primary production, and pigment analysis). Water samples were taken from three replicate bottles under each condition and used for the measurement of dissolved oxygen and alkalinity. The same experimental design was used for individual bacterial testing. Corals were incubated for 4 days at 32 °C in seawater enriched with bacteria at a final total abundance of 10^6 cells ml^{-1}. Five bottles containing one branch each were prepared for each bacterium. The final state of the corals was evaluated by their coloration, their maximum photochemical yield (Fv/Fm), and the presence or absence of tissue necrosis

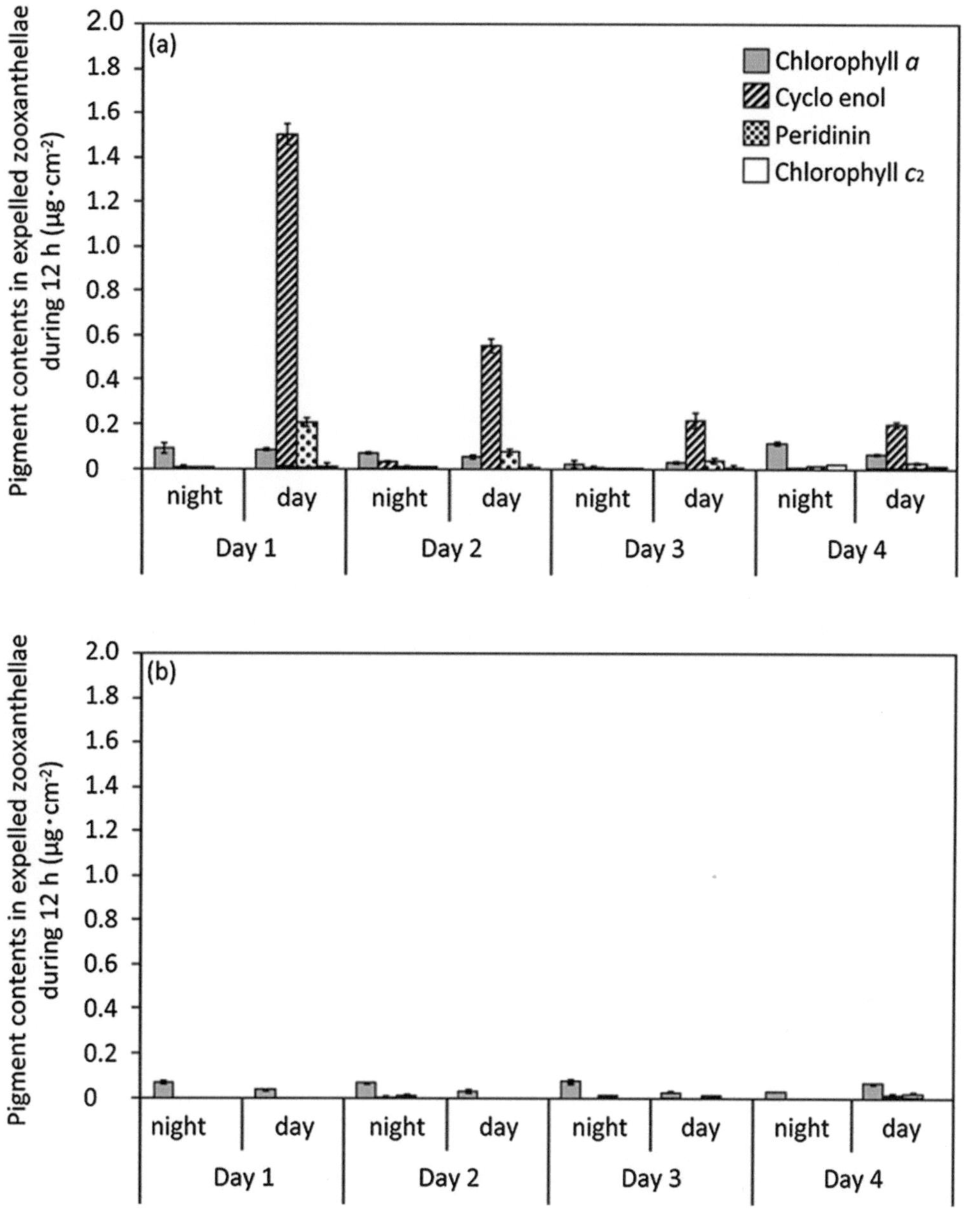

Fig. 2.6 Pigments compositions of zooxanthellae expelled from coral branches held under a 12-h photoperiod over 4 days of incubation. **a** 27 °C and **b** 32 °C. Values are means±SD (n = 3) (Reproduced from Suzuki et al. (2015) by permission of John Wiley & Sons Ltd. Copyright © 2014 The Authors. *Journal of Phycology* published by Wiley Periodicals, Inc. on behalf of Phycological Society of America)

2.2.3 Effect of Bacteria and Their Role in Coral Metabolisms

After 8 days of incubation, the loss of zooxanthellae under high temperature was clearly visible especially for those corals under thermal stress. Zooxanthellae density of *M. digitata* did not significantly decrease with the addition of bacteria at 27 °C; however, corals were bleached dramatically by synergistic stresses of high temperature and bacteria. *M. digitata* lost about 45 % of its zooxanthellae under high temperature and about 70 % under combined high temperature and bacteria on day 8 of incubation (Fig. 2.9a–f). This means that combination of thermal stress and bacteria strongly influenced coral bleaching. In terms of chlorophyll *a*, in controls at initial time, it was 2.0 μg cm^{-2} and slightly increased to 2.4 μg cm^{-2} in controls after 8 days of incubations; however, it decreased to 1.7 μg cm^{-2} at 27 °C with addition of bacteria. At stress temperature (32 °C), it decreased to 1.7 μg cm^{-2} and 1.3 μg cm^{-2} in combined stresses of 32 °C and bacteria addition. Moreover chlorophyll *a* content per zooxanthellae varied from 5.2 picograms chlorophyll *a* cell^{-1} to 4.6 picograms cell^{-1} in control after

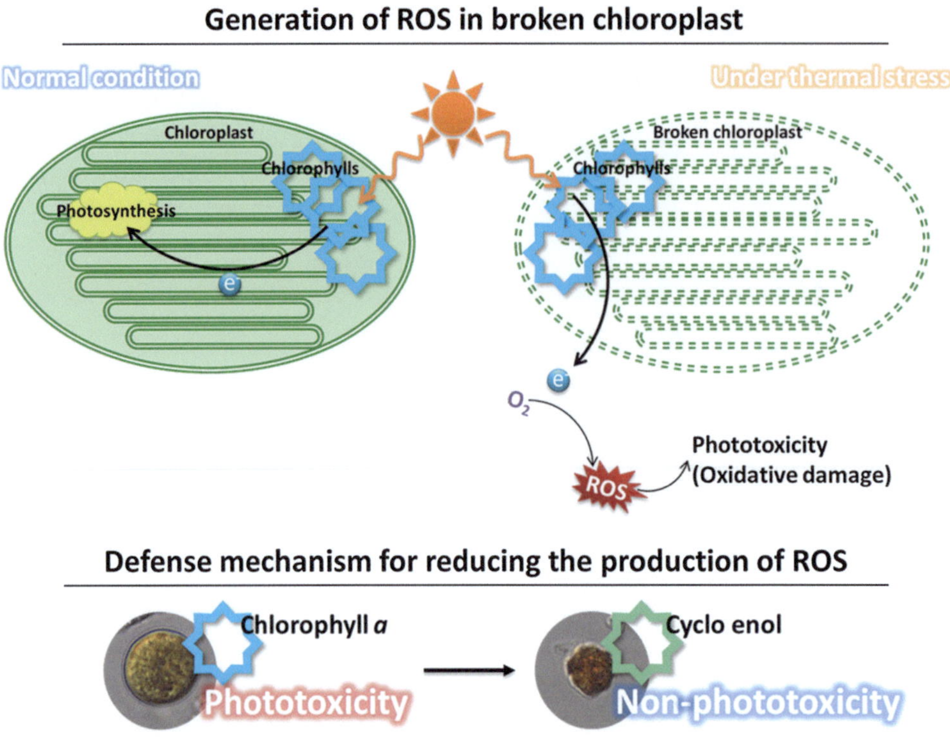

Fig. 2.7 Generation of ROS in broken chloroplast and significance of the conversion of chlorophyll *a* to cyclo enol

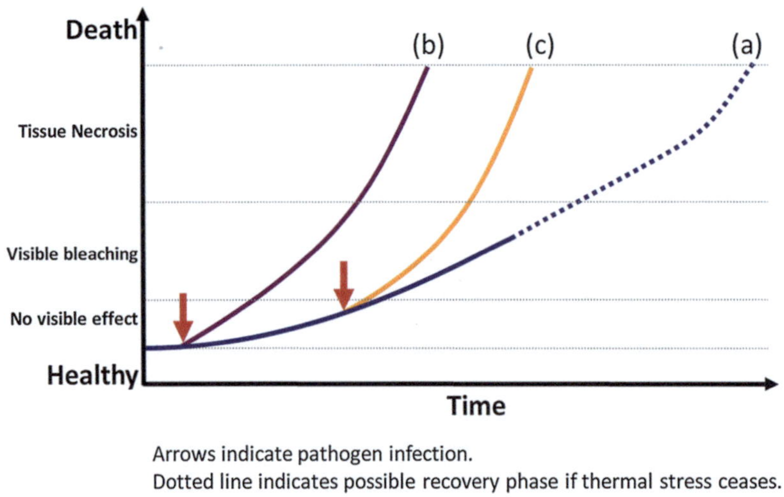

Fig. 2.8 Three possible ways of coral bleaching

8 days of incubations and 3.8 picograms cell^{-1} at 27 °C with addition of bacteria, but it increased to 6.1 picograms cell^{-1} at high temperature treatment and to 8.5 picograms cell^{-1} in combined stresses treatment. This may indicate that corals retain healthy zooxanthellae that contain intact and replete chloroplasts in order to face disadvantages during bleaching events (Fig. 2.9d, e). The *Fv*/*Fm* ratio (Fig. 2.9f) significantly changed at high temperature with bacterial addition (HSD, $P < 0.05$). A two-way ANOVA showed a significant effect of temperature on the maximum photosynthetic yield on day 8 ($P = 0.003$). The *Fv*/*Fm* ratio of coral at high temperature with inoculation was significantly lower than the ratios of corals with the other treatments, with average values of 0.57 ± 0.02 and 0.67 ± 0.05 and 0.77 ± 0.02 and 0.74 ± 0.04 at 32 °C and 27 °C with and without bacteria, respectively. Thus, bacterial inoculation decreased photosynthetic efficiency by about 25 % at 32 °C.

At the same time, synergistic stress of high temperature with bacteria inoculation significantly affected primary production rates (Fig. 2.10a): at elevated temperature (32 °C), primary

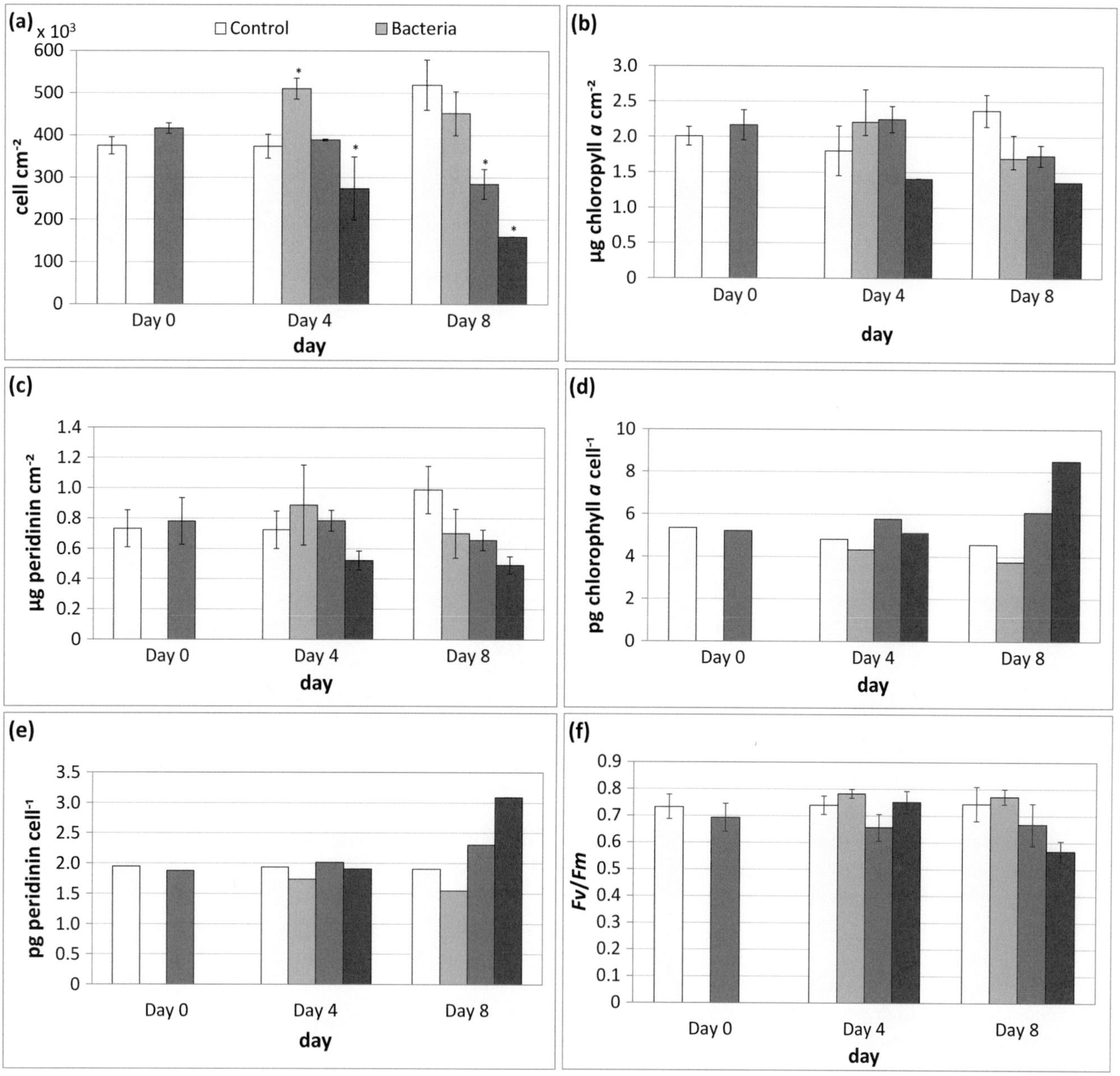

Fig. 2.9 Variation in (**a**) zooxanthellae density in coral tissue, (**b**) chlorophyll *a* content, (**c**) peridinin content, (**d**) chlorophyll *a* per zooxanthellar cell, (**e**) peridinin per zooxanthellar cell, and (**f**) maximum quantum yield (*Fv*/*Fm*) with two levels of bacteria abundance (no addition and addition of five bacterial strains) at ambient (27 °C) and at increased temperature (32 °C). Values are mean values±SE ($n = 3$ for each treatment). *$P < 0.05$ compared to the control according to the Tukey–Kramer HSD test

production of the control (1.03 ± 0.06 μg C cm^{-2} h^{-1}) decreased 28 % to an average value of 0.74 ± 0.04 μg C cm^{-2} h^{-1} (HSD, $P < 0.05$), and with synergistic effects of high temperature and bacteria addition, primary production was also significantly decreased by about 66 % (HSD, $P < 0.05$), to an average of 0.35 ± 0.10 μg C cm^{-2} h^{-1}, while respiration rates (Fig. 2.10b) increased by approximately 30 % under high temperature to reach a value of 0.50 ± 0.03 and 0.49 ± 0.01 μmol O_2 cm^{-2} h^{-1} for high temperature without and with bacteria, respectively. However, addition of bacteria did not affect respiration rates at both high temperature and normal temperature. Respiration, which reflects host metabolism and to a lesser extent zooxanthellae metabolism, followed a similar pattern as photosynthesis variation but with a smaller amplitude and reverse direction. The higher respiration rates shown here may be attributable to higher energy requirements to repair the damage caused by temperature stress.

The calcification rate (Fig. 2.10c) was not significantly decreased by the addition of bacteria at 27 °C, with rates of

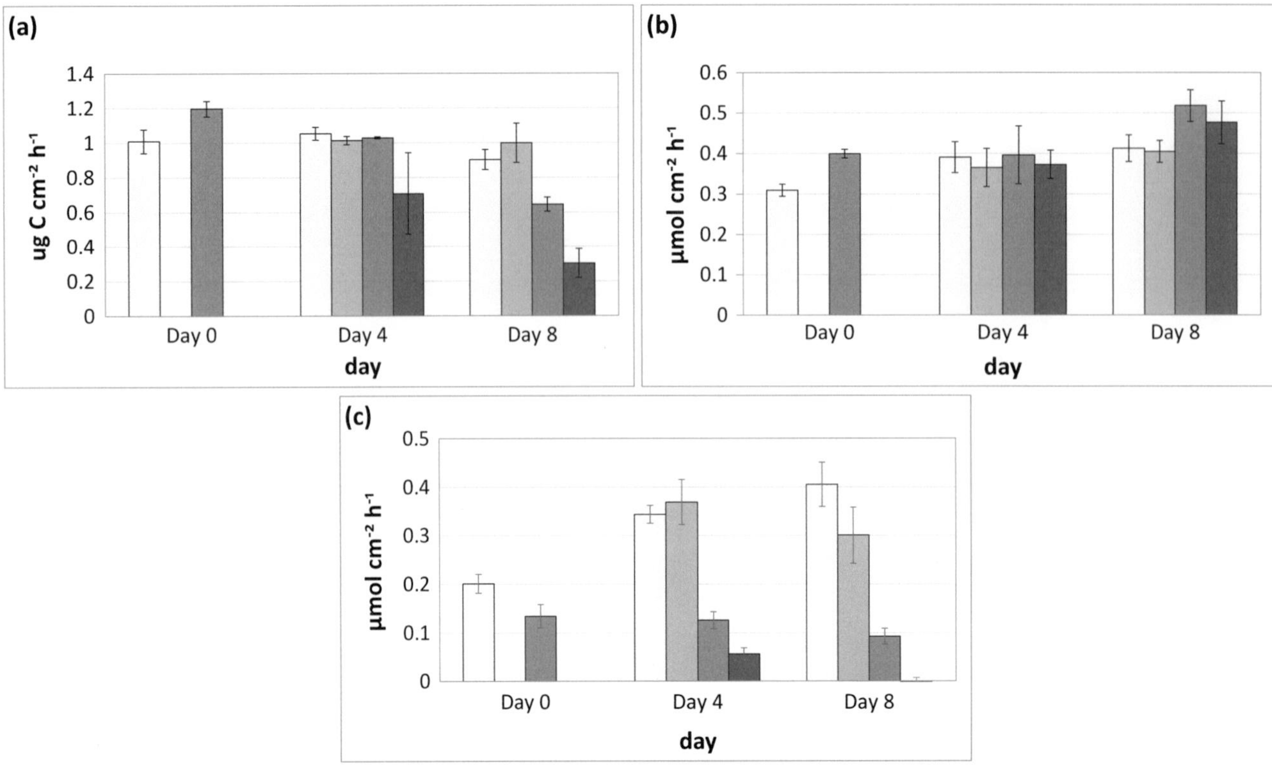

Fig. 2.10 Variation in (**a**) primary production, (**b**) respiration, and (**c**) calcification with two levels of bacteria abundance (no addition and addition of five bacterial strains) at ambient (27 °C) and at increased temperature (32 °C). Values are mean values±SE ($n = 3$ for each treatment). $*P < 0.05$ compared to the control according to the Tukey–Kramer HSD test

0.41 ± 0.05 µmol cm^{-2} h^{-1} for the control treatment and 0.30 ± 0.06 µmol cm^{-2} h^{-1} with the addition of bacteria. However, temperature stress significantly affected the calcification rate (ANOVA, $P < 0.001$), which decreased to 0.09 ± 0.02 µmol cm^{-2} h^{-1} at 32 °C, representing a decrease of about 77 % (HSD, $P < 0.05$). Moreover, the combined stresses of high temperature and bacteria greatly decreased the calcification rate, by about 101 % (HSD, $P < 0.05$), with a net dissolution of calcium carbonate from the coral occurring on day 8 at 32 °C with bacteria (-0.01 ± 0.01 µmol cm^{-2} h^{-1}).

During a challenge with individual bacterial strains, the appearance of branches that were incubated in the presence of *V. coralliilyticus*, *V. harveyi*, *P. carotinifaciens*, or *Pseudoalteromonas* sp. did not change after 4 days of incubation at 32 °C and appeared to be as healthy as the branches shown in Fig. 2.11a, b. The *Fv*/*Fm* ratios of these branches were relatively stable. All five replicates maintained their coloration and expanded their polyps. However, branches in the presence of *Sulfitobacter* sp. at 32 °C exhibited signs of severe bleaching and some tissue necrosis (Fig. 2.11e), and the *Fv*/*Fm* ratio was decreased by about 24 % compared with control branches.

2.2.4 Significance of Synergistic Actions of Thermal Stress and Pathogens Inputs

Individually, only *Sulfitobacter* sp. provoked acute bleaching of the coral after 4 days of incubation at 32 °C. The other strains did not cause visible effects on *M. digitata* in these conditions. However, *V. coralliilyticus* (Ben-Haim et al. 2003; Sussman et al. 2008), *P. carotinifaciens* (Casareto, pers. comm.), and *V. harveyi* (Sutherland et al. 2004; Gomez-Gil et al. 2004) were reported as coral pathogens. These three last strains individually, without a trigger other than temperature, may not be responsible for the bleaching of *M. digitata*. Lesions on corals under thermal stress with a mixture of the five bacteria and under thermal stress with *Sulfitobacter* sp. alone showed a loss of coloration and some tissue necrosis.

With *Sulfitobacter* sp. alone, the lesions were more severe. At high temperature without the addition of bacteria, moderate loss of coloration was observed, but there were no signs of necrosis. Tissue necrosis can be interpreted as a more advanced compromised stage following bleaching. Five mixed bacteria did not have any effects at 27 °C. Therefore, *Sulfitobacter* sp. or these bacteria when mixed

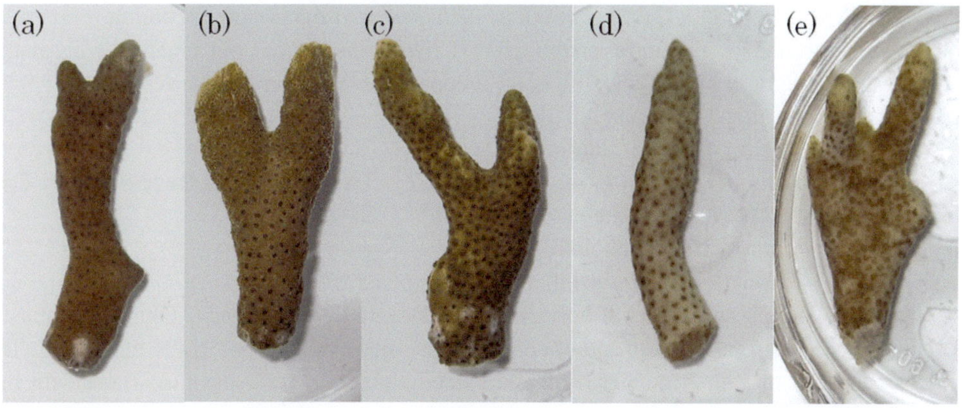

Fig. 2.11 Photograph of branches of *Montipora digitata* after incubation under each condition. (**a**) At 27 °C without addition of bacteria, (**b**) at 27 °C with addition of five mixed bacteria (*Vibrio coralliilyticus*, *Vibrio harveyi*, *Paracoccus carotinifaciens*, *Pseudoalteromonas* sp., and *Sulfitobacter* sp.), (**c**) at 32 °C without addition of bacteria, (**d**) at 32 °C with addition of five mixed bacteria, and **e** at 32 °C with addition of *Sulfitobacter* sp.

required a temperature higher than normal to cause damage to *M. digitata*. This phenomenon may be due to two different mechanisms: the bacteria express some virulence factor only at high temperature as for the *Oculina patagonica*/*V. shiloi* model (Rosenberg and Falkovitz 2004), or the coral immune system was weakened by the heat stress allowing the action by these bacteria (Mylardz et al. 2009). The effect of bacteria as accelerator for coral bleaching is large with high temperature.

Over a short period, moderate thermal stress may not have a strong effect on corals, and if temperatures return to normal levels, it is highly probable that the corals would be able to recover. Under normal temperature, bacteria did not affect coral bleaching and metabolism. However, under high temperature, bacteria caused severe bleaching and decreased photosynthesis and calcification activity, suggesting that potentially pathogenic bacteria can have a dramatic negative impact on corals under thermal stress. However, respiration did not increase with the addition of bacteria under high temperature. The tested bacteria, especially *Sulfitobacter* sp., greatly enhanced and accelerated the bleaching process. These bacteria were isolated from the seawater surrounding the bleached corals and from the corals themselves. They can be found in the seawater around Okinawa. If their populations increase due to still undetermined factors, the damage to coral reefs in this area, which are under thermal stress with increasing frequency, could be dramatic.

2.2.5 Summary of Sect. 2.2

In this research the hypothesis that coral bleaching triggered by thermal stress disrupts coral metabolism, weakening the coral and allowing bacterial challenges to affect the host, was tested. This secondary process of bacterial challenges enhances and accelerates the bleaching process, which involves the loss of zooxanthellae and changes in metabolic processes. Such successional unfavorable events may become quite common under the combined effects of various environmental changes linked to climate change as higher seawater temperature and changes in water quality increasing the number of opportunistic pathogens in reef waters. The hypothesis was tested in the laboratory using the coral *Montipora digitata* and challenges from five bacterial species (*Vibrio coralliilyticus*, *Vibrio harveyi*, *Paracoccus carotinifaciens*, *Pseudoalteromonas* sp., and *Sulfitobacter* sp.) previously isolated from corals and their surrounding seawater.

Bacterial challenges in addition to high-temperature stress resulted in coral bleaching, with a 70 % decrease in zooxanthellae density compared with the control, a 25 % decrease in photosynthetic efficiency (*Fv*/*Fm*), a 66 % decrease in photosynthesis, and a 101 % reduction in calcification activity. Tissue necrosis was observed in the most compromised branches. Among the bacteria examined, *Sulfitobacter* sp. had a greater capacity to enhance and accelerate the bleaching process under thermal stress.

2.3 Synergistic Effects of Thermal Stress and High Nutrient Levels on the Scleractinian Coral *Pocillopora damicornis*

2.3.1 Introduction

Coral reefs, the most diverse marine ecosystem, are known to harbor a wide variety of marine organisms providing important goods and services to coastal communities through income generation such as tourism, fishing, and building materials (Hoegh-Guldberg et al. 2007; Hughes et al. 2003). Coral reefs support high production even though nutrient levels are very low (oligotrophic). Typically, coral reef waters consist of 0.6 μM of nitrate (NO_3) and 0.2 μM of phosphate (PO_4), which are the main sources of nutrient for organism surviving in these waters (Kleypas 1994). The success of reef-building corals in this oligotrophic environment is attributed to the symbiosis with phototrophic dinoflagellates commonly known as zooxanthellae (Muscatine 1990). In this relationship, the coral host gets photosynthetic organic products from the zooxanthellae to fuel important processes like respiration, tissue growth, and calcification (Muscatine 1990). Disruption of this host–symbiont relationship eventually leads to coral bleaching by loss of algal cells and/or algal pigments (Douglas 2003; Hoegh-Guldberg 1999; Hughes et al. 2003; Suzuki et al. 2015). Several environmental factors like increased sea surface temperature, high light, low water quality, and bacterial infections have been found to be the cause of coral bleaching (Douglas 2003; Fitt et al. 2001; Higuchi et al. 2013; Hoegh-Guldberg 1999). In addition to these factors, excess input of nutrients in the marine ecosystem, due to sewage dumping, pollution, and runoff from agricultural areas, has been shifting the composition of coral reefs by favoring the outgrowth of macroalgae (Birkeland 1987; Fabricius and De'ath 2004; Stimson et al. 2001). However, the effects of high nutrient concentration in combination with higher than normal seawater temperature on corals are still not fully understood due to the complexity of the host–symbiont relationship. Studies showing no adverse effect on coral physiology in regions of high nutrient level, in addition with reports of increase in zooxanthellae cell density due to availability of high external nitrogen level, may seem to indicate that corals can survive in high nutrient areas (Agostini et al. 2012; Atkinson et al. 1995; Bongiorni et al. 2003; Fabricius 2005; Marubini and Davies 1996; Muscatine et al. 1989; Szmant 2002). However, recent studies have shown that anthropogenic nutrient input into the ocean can result in nutrient imbalance and cause unavailability of other nutrients like phosphate (Fabricius 2005; Parkhill et al. 2001; Wiedenmann et al. 2012). This may lead to phosphate starvation of the symbiont, causing a decrease in maximum quantum yield (*Fv*/*Fm*) and thus making the coral host more susceptible to bleaching (Parkhill et al. 2001; Wiedenmann et al. 2012).

Even though respond of corals to high temperature and high nitrate concentration has been widely researched, very little is known about recovery of these corals after the stress exposure. So, this study was carried out to investigate the effects of high temperature and high nitrate concentration on the coral *Pocillopora damicornis* and the ability of the coral holobiont to recover from these stresses. During the incubation experiment, measurement of maximum quantum yield (*Fv*/*Fm*) and maximum excitation pressure on photosystem II (*Qm*) was carried out together with analysis of pigment content of the zooxanthellae.

2.3.2 Testing Coral Response to Combined Stresses

Fragments of *Pocillopora damicornis* ($n = 32$) were sampled from Sesoko Beach, Okinawa, Japan (26°38′N 127° 51′E) at a depth of 1–3 m during low tide. After acclimatization for 3 days in a tank with running seawater and ambient light conditions, the coral nubbins were moved to the incubation system. The collected nubbins were placed in 800 ml containers at ambient temperature. The containers were supplied with filtered seawater by peristaltic pumps, and the use of stirrers provided water movement inside each container (Fujimura et al. 2008).

Incubation of *Pocillopora damicornis* ($n = 2$) with different seawater temperature and nitrate conditions was carried out so as to investigate the combined effects of high temperature and nitrate level. High temperature and high nitrate stress (HN32) was carried out by subjecting the nubbins to 32 °C and providing a constant supply of filtered seawater (0.2 μm) with a nitrate concentration of 10 μmol/L. The control (AN27) was kept at 27 °C and constantly supplied with ambient nitrate filtered seawater (without addition of nitrate). Conditions to evaluate the effect of high nitrate concentration (HN27) and high temperature (AN32) separately were also set up. The light level was kept at 200μmolm^{-2}s^{-1} and followed a daily routine of 12-h light and 12-h dark cycle. High temperature and nitrate stress were carried out for 2 days followed by 2 days of recovery at ambient temperature and nitrate concentration. For evaluation on the effect of high nutrients under the thermal stress to the coral symbiont, measurement for chemical and biological parameters, examples PAM data, chlorophyll *a*, pigments, and zooxanthellae density.

2.3.3 Effect of High Nitrate Concentrations Under the Thermal Stress

Variation in photochemical efficiency of PSII (*Fv*/*Fm*) throughout the experiment indicated photophysiological stress in zooxanthellae of *P. damicornis* and subsequent recovery of the photosynthetic system of the zooxanthellae in different incubation conditions. The coral fragments that were incubated under high temperature (AN32), high nitrate (HN27), and high nitrate combined with high temperature (HN32) showed a significant decrease in *Fv*/*Fm* after 48 hrs of stress phase when compared with the control (AN27). During the first day of recovery, only corals under HN27

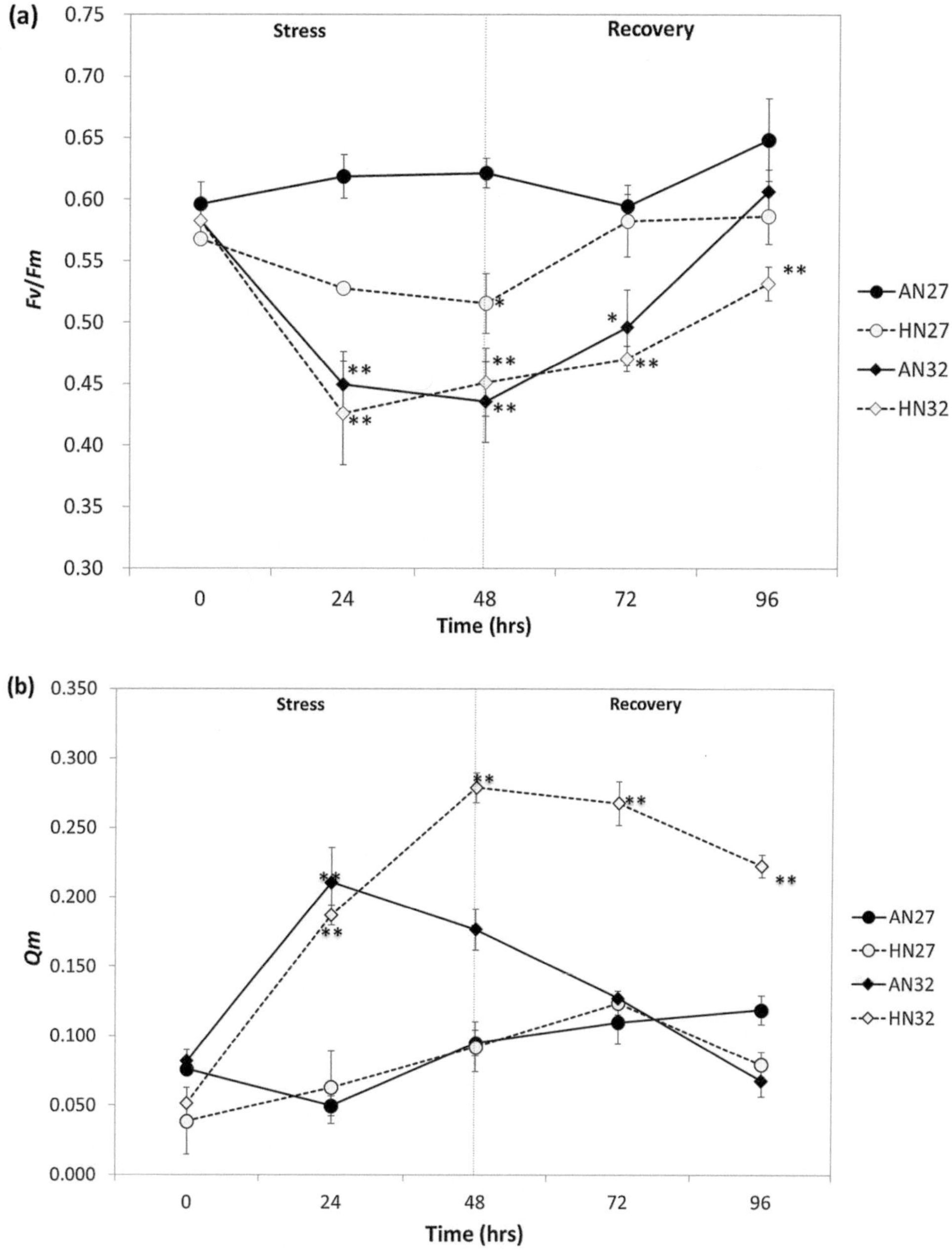

Fig. 2.12 (**a**) Changes in maximum quantum yield of corals during stress and recovery phases. (**b**) Changes in maximum excitation pressure on PSII of corals during stress and recovery phases. Different nitrate and temperature conditions were used during the incubation: ambient nitrate at 27 °C (AN27), high nitrate at 27 °C (HN27), ambient nitrate at 32 °C (AN32), and high nitrate at 32 °C (HN32). A three-way ANOVA test was performed and Tukey HSD post hoc test was then applied to the data for comparison with the control (AN27). Asterisks represent significant differences ($*p < 0.05$, $**p < 0.01$). Vertical bars represent standard deviation from mean value (n = 2)

showed almost complete recovery, whereas corals under AN32 and HN32 still had significantly lower *Fv/Fm* values. After 2 days of recovery phase, corals under all the conditions showed good recovery in *Fv/Fm*, except those subjected to HN32. For these corals, *Fv/Fm* (0.522 ± 0.014; $n = 2$) was significantly lower than the control (0.625 ± 0.033; $n = 2$) even after 48 h of recovery (Fig. 2.12).

Zooxanthellae density varied in the different incubation conditions after stress and recovery phases. After a 2-day stress period, zooxanthellae density of corals in HN27, AN32, and HN32 was significantly lower than the control (AN27) (Fig. 2.13). As for the recovery phase, an increase in zooxanthellae cell density was noted for corals in AN32 and HN32 with a significantly higher zooxanthellae cell density for HN32 when compared to the control (Fig. 2.13). When

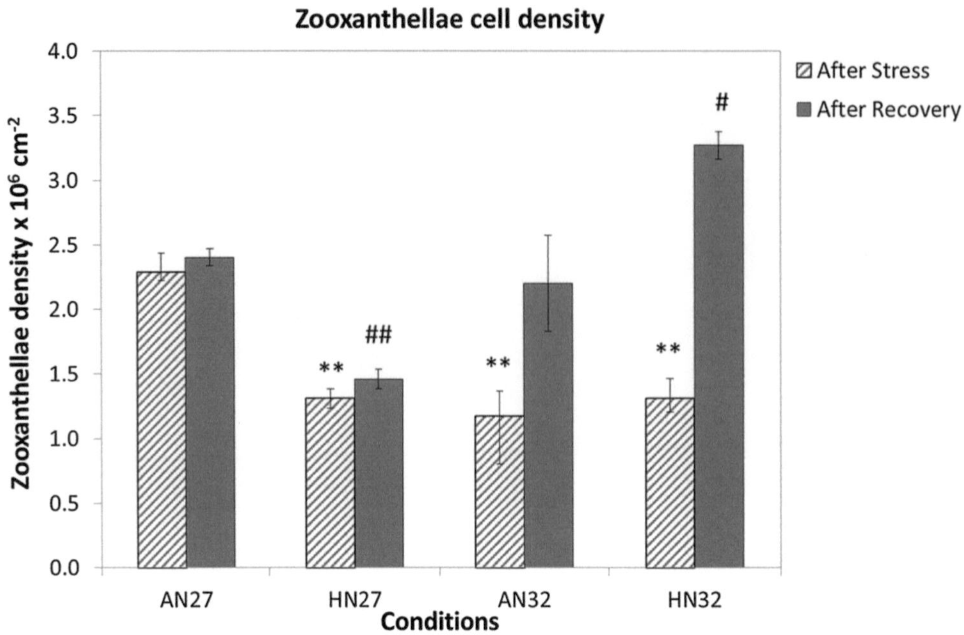

Fig. 2.13 Zooxanthellae cell density after stress and recovery periods. Corals incubated in ambient nitrate (AN27), high nitrate (HN27), high temperature (AN32), and a combination of high nitrate and high temperature (HN32) were tested statistically for significant differences when compared to control (AN27). A three-way ANOVA test and Tukey HSD post hoc test were used. * and # represent significant differences with the control for stress and recovery phases, respectively ($*p < 0.05$, $*p < 0.01$; $\#p < 0.05$, $\#\#p < 0.01$). Vertical bars represent standard deviation from mean value ($n = 2$)

comparison was made with zooxanthellae density from after stress period, corals incubated in AN32 and HN32 showed an 88 % and 150 % increase, respectively. Corals in high nitrate condition (HN27), after the recovery phase, still had significantly lower zooxanthellae cells than the control and showed only an 11 % increase from the stress phase.

Zooxanthellae cells possess both chlorophyll *a* and peridinin as photosynthetic pigments. Contrary to chlorophyll *a*, which is also found in epilithic and endolithic algae associated with the coral, peridinin is specific to zooxanthellae cells and thus gives a more accurate indication of pigment changes in zooxanthellae cells. Chlorophyll *a* and peridinin content per cell in corals during the whole incubation experiment followed a similar trend. After the stress phase, no significant change was observed in chlorophyll *a* and peridinin contents (Fig. 2.14a, b) in all the conditions. After recovery, when compared to the control, both chlorophyll *a* and peridinin per cell decreased significantly for corals in HN32. For this condition, the chlorophyll *a* and peridinin content decreased by 44 % and 46 %, respectively, when compared with the stress phase.

2.3.4 Stress Phase

During the stress phase, zooxanthellae density decreased for all the condition compared to the control (Fig. 2.13). Even though zooxanthellae cell decreased, it was noted that chlorophyll *a* and peridinin concentration per cell was relatively unaffected (Fig. 2.14a, b). This indicated that the corals under nitrate or temperature stress may have released or digested zooxanthellae cells that have been damaged and cannot function properly. This mechanism prevented the formation of excess reactive oxygen species (ROS) by "unhealthy" zooxanthellae cells that can damage the host and remaining zooxanthellae (Lesser and Farrell 2004; Warner et al. 1999; Suzuki et al. 2014). Compared to other studies carried out, an increase in zooxanthellae cell density was not observed after incubation with high nitrate concentration only (Fabricius 2005; Miller and Yellowlees 1989). Nutrient imbalance due to the availability of abundant nitrate could be the cause of this decrease in zooxanthellae density. Under such case, phosphate usually becomes the limiting factor for growth and proliferation (Miller and Yellowlees 1989). Recent studies have shown that under phosphate deficiency, the anionic environment of the thylakoid membrane is disrupted by the excessive conversion of the phospholipid phosphatidylglycerol (PG) to the sulpholipid sulfoquinovosyldiacylglycerol (SQDG) (Frentzen 2004; Wiedenmann et al. 2012). This could explain the decrease of zooxanthellae cells for corals under high nitrate (Fig. 2.13).

Fv/*Fm* and *Qm* values throughout the experiment tend to indicate that the combined effect of high temperature and

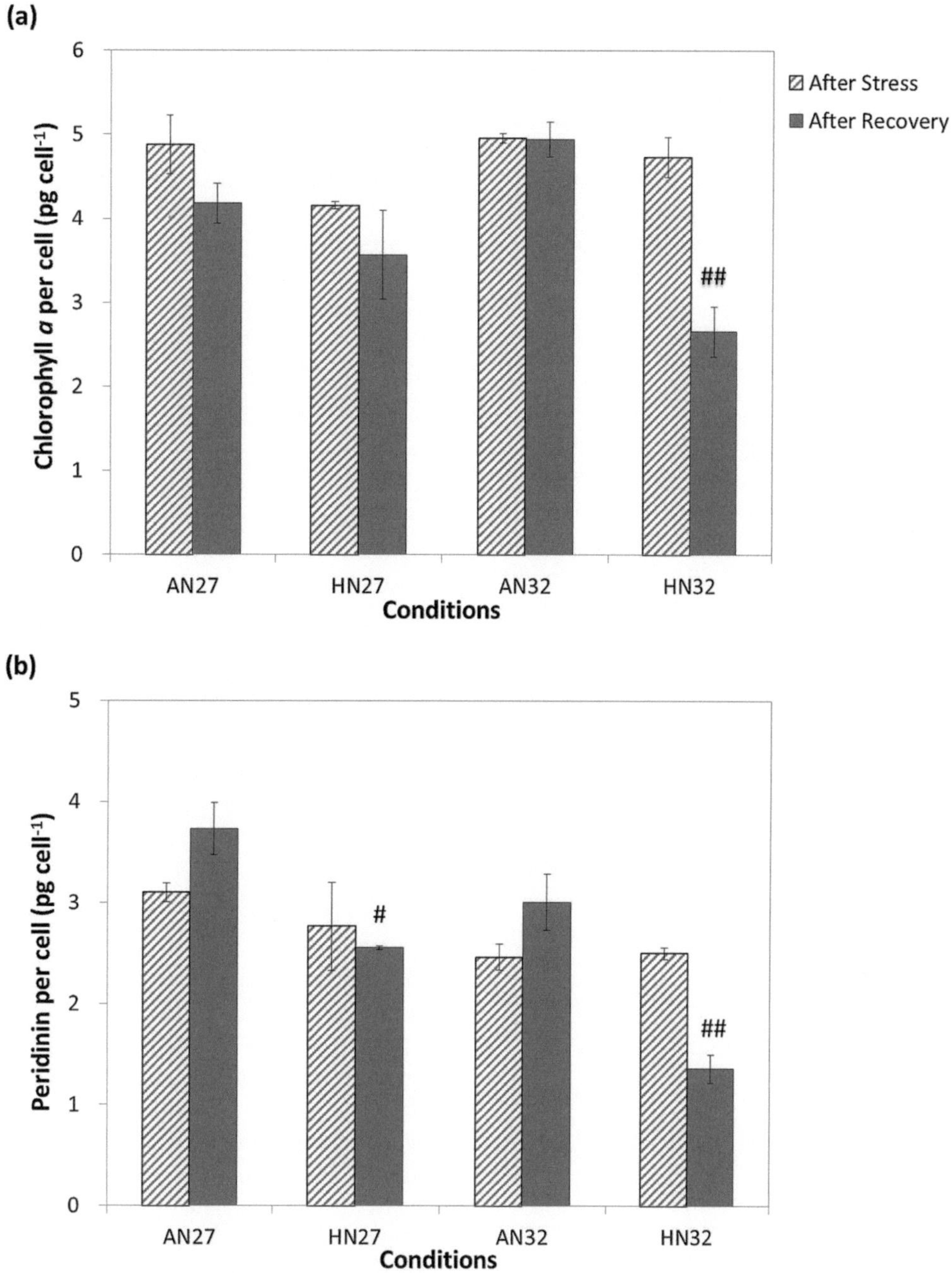

Fig. 2.14 Pigment content in zooxanthellae cells after stress and recovery periods. Chlorophyll *a* per cell **a** and peridinin per cell **b** were monitored for different incubation conditions: ambient nitrate (AN27), high nitrate (HN27), high temperature (AN32), and high nitrate together with high temperature (HN32). A three-way ANOVA test and Tukey HSD post hoc test were used to distinguish significant changes in pigment contents. # shows significant differences with the control for recovery period. No significant differences were found for after stress period (#$p < 0.05$, ##$p < 0.01$). Vertical bars represent standard deviation from mean value ($n = 2$)

high nitrate concentration may produce a more severe stress than high temperature alone (Fig. 2.12a, b). The decrease in Fv/Fm for both AN32 and HN32 can be attributed to the increase in ROS due to photoinhibition that can be harmful to both host and symbiont (Lesser and Farrell 2004), but higher Qm for HN32 after the stress phase may indicate that high nitrate can enhance temperature-mediated stress. In case of HN32, where the coral was already under thermal stress, high concentration of NO_3 could instigate production of nitric oxide (NO). Previous studies carried out on the plant enzyme nitrate reductase showed that this enzyme could reduce NO_3^- to NO, and in addition to this, NO synthesis by nitric oxide synthetase (NOS) in the coral holobiont was found to increase during thermal stress (Perez and Weis 2006; Trapido-Rosenthal et al. 2005; Yamasaki and Sakihama 2000). So, NO_3^- uptaken by zooxanthellae cells can be reduced to nitrite (NO_2^-) and finally to NO through enzymatic reactions. Even though NO is a natural signaling molecule in animals, it also acts as a cytotoxic molecule. In the presence of ROS, the highly reactive peroxynitrite ($ONOO^-$) can be produced from NO (Weis 2008). The combination of ROS, NO, and $ONOO^-$ could cause damaged to the thylakoid membrane of zooxanthellae, thus increasing *Qm* and decreasing *Fv*/*Fm* for HN32 during the stress phase.

2.3.5 Recovery Phase

Response of corals during the recovery phase varied greatly according to the conditions they experienced during the stress phase. Zooxanthellae density among recovering corals showed an increase for AN32 and HN32, but the pigment contents did not show a similar trend. In case of AN32, the increase in zooxanthellae cells was accompanied with an unchanged concentration of chlorophyll *a* and peridinin per cell, whereas for HN32, a decrease in chlorophyll *a* and peridinin per cell was noted with increasing zooxanthellae density. This decrease in chlorophyll *a* content in zooxanthellae cells for HN32 could be the reason for the partial recovery of *Fv*/*Fm* compared to AN32 that showed almost complete recovery. Even though the zooxanthellae cell density increased more than the control, a below-normal pigment content could have hampered the recovery of corals that experienced both high nitrate concentration and high temperature.

Thermally stressed corals recovered by selective degradation and replacement of damaged proteins to repair affected PSII systems. As photosynthetic efficiency is restored, the coral is able to overcome the effects that occurred due to the stress. Increase in photosynthesis efficiency will allow production of more photosynthetic products by the zooxanthellae that will be used for restoration of damaged PSII sites and proliferation of healthy zooxanthellae cells. These energy-rich products, such as glycerol and glucose, are then translocated to the coral and used by the latter for respiration (Muscatine et al. 1984; Trench 1993). This is the case with AN32 where good recovery in *Fv*/*Fm* was observed. On the other hand, data obtained in this experiment may indicate that the damage caused during high nitrate and high temperature can impede the recovery of these corals. The combined effects of ROS, NO, and $ONOO^-$ could affect not only the thylakoid membrane but also directly degrade DNA content of both host and symbiont which may result in cell apoptosis (Pacher et al. 2007; Weis 2008). So, during the recovery phase, above normal zooxanthellae proliferation could occur to compensate for the damage. High density of low chlorophyll *a* content cells may help the coral to overcome the effects of the stress phase, and when it achieves constant photosynthetic efficiency, these cells are replaced by healthier ones.

This study demonstrated that an increase in nitrate and temperature could be harmful to the coral *Pocillopora damicornis*. Also recovery can be more difficult due to the severity of damage that occurred during the stress period. Short and sudden increases in nitrate and temperature levels can impede recovery of the coral so that they are more susceptible to other environmental or anthropogenic stresses. Care should be taken when judging the impacts of high nitrate and high temperature as acclimatization to these new environmental conditions may require longer exposure time in further experiments.

2.3.6 Summary for Sect. 2.3

Increase in sea surface temperature and excessive input of nutrient in reef waters, due to anthropogenic activities, have been among the various factors responsible for bleaching and mortality of corals around the world. Moreover synergistic effects of these environmental stressors can accelerate and aggravate bleaching process and infections of possible opportunistic pathogens that take advantage of these environmental changes.

In this study the effects of elevated nitrate (NO_3) concentration in combination with high seawater temperature on the coral, *Pocillopora damicornis*, and the ability of the coral holobiont to recover from these stresses were investigated. Coral fragments were incubated for 2 days at different temperatures (27 °C and 32 °C) and NO_3 concentrations (<1 μmol/L and 10 μmol/L). Following 2 days of stress under 32 °C and 10 μmol/L NO_3, the nubbins were moved to 27 °C and ambient (<1 μmol/L) NO_3 levels for 2 days of recovery period. Maximum quantum yield (*Fv*/*Fm*) and maximum excitation pressure (*Qm*) at photosystem II indicated that combined effects of high temperature and high NO_3 were more severe and only corals under these conditions exhibited an incomplete recovery. Furthermore, zooxanthellae density and pigment data showed that the response mechanism of these coral nubbins was different from high temperature or high NO_3 stress only. During the recovery phase, zooxanthellae density was higher (~0.5-fold), whereas chlorophyll *a* per cell was lower (~0.5-fold) than the control for combined stress of high nitrate and temperature. Hence, this study showed that under thermal stress, high nitrate amplifies damage to the zooxanthellae of the coral *Pocillopora damicornis* and recovery of the holobiont is more difficult after the stress.

Acknowledgments We would like to express our sincere thanks to Dr. Hiroyuki Fujimura of the University of the Ryukyus, Dr. Tomihiko Higuchi of Shizuoka University, and Dr. Pramod K. Chumun of

Mauritius University for cooperative research. The project was supported by a Grant-in-Aid for Scientific Research on Innovative Areas for "Coral reef science for symbiosis and coexistence of human and ecosystem under combined stresses" (20121003) of the Ministry of Education, Culture, Sports, Science and Technology (MEXT), Japan, and the Global Coral Reef Conservation Project (GCRCP) of Mitsubishi Corporation, Japan. We are also thankful to Tropical Biosphere Research Center of the University of the Ryukyus at Sesoko Island, Okinawa, Japan, for providing laboratory facilities.

References

Agostini S, Suzuki Y, Casareto BE, Nakano Y, Hidaka M, Badrun N (2009) Coral symbiotic complex: hypothesis through vitamin B12 for a new evaluation. Galaxea JCRS 11:1–11

Agostini S, Suzuki Y, Higuchi T, Casareto BE, Yoshinaga K, Nakano Y, Fujimura H (2012) Biological and chemical characteristics of the coral gastric cavity. Coral Reefs 31 (1):147–156

Atkinson MJ, Carlson B, Crow GL (1995) Coral growth in high-nutrient, low-pH seawater: a case study of corals cultured at the Waikiki Aquarium, Honolulu, Hawaii. Coral Reefs 14:215–223

Ben-Haim Y, Zicherman-Keren M, Rosenberg E (2003) Temperature-regulated bleaching and lysis of the coral *Pocillopora damicornis* by the novel pathogen *Vibrio coralliilyticus*. Appl Environ Microbiol 69:4236–4242

Bhagooli R, Hidaka M (2002) Physiological responses of the coral *Galaxea fascicularis* and its algal symbiont to elevated temperatures. Galaxea JCRS 4:33–42

Bhagooli R, Hidaka M (2003) Comparison of stress susceptibility of in hospite and isolated zooxanthellae among five coral species. J Exp Mar Biol Ecol 291:181–197

Bhagooli R, Hidaka M (2006) Thermal inhibition and recovery of the maximum quantum yield of photosystem II and the maximum electron transport rate in zooxanthellae of a reef-building coral. Galaxea JCRS 8:1–11

Birkeland C (1987) Nutrient availability as a major determinant of differences among coastal hard-substratum communities in different regions of the tropics. In: Comparison between Atlantic and Pacific tropical marine coastal ecosystems: community structure, ecological processes, and productivity, UNESCO Reports in Marine Science. UNESCO, Paris, pp 45–97

Bongiorni L, Shafir S, Angel D, Rinkevich B (2003) Survival, growth and gonad development of two hermatypic corals subjected to in situ fish-farm nutrient enrichment. Mar Ecol Prog Ser 253:137–144

Bourne D, Iida Y, Uthicke S, Smith-Keune C (2008) Changes in coral-associated microbial communities during a bleaching event. ISME J 2:350–363

Brown BE, Le Tessier MDA, Bythell JC (1995) Mechanisms of bleaching deduced from histological studies of reef corals sampled during a natural bleaching event. Mar Biol 122:655–663

Chillier XFD, Gulacar FO, Buchs A (1993) A novel sedimentary lacustrine chlorin: characterization and geochemical significance. Chemosphere 27:2103–2110

Douglas AE (2003) Coral bleaching—how and why? Mar Pollut Bull 46:385–392

Downs CA, Fauth JE, Halas JC, Dustan P, Bemiss J, Woodley CM (2002) Oxidative stress and seasonal coral bleaching. Free Radic Biol Med 33:533–543

Downs CA, Kramarsky-Winter E, Martinez J, Kushmaro A, Woodley CM, Loya Y, Ostrander GK (2009) Symbiophagy as a cellular mechanism for coral bleaching. Autophagy 5(2):211–216

Downs CA, McDougall KE, Woodley CM, Fauth JE, Richmond RH, Kushmaro A, Gibb SW, Loya Y, Ostrander GK, Kramarsky-Winter E (2013) Heat-stress and light-stress induce different cellular pathologies in the symbiotic dinoflagellate during coral bleaching. PLoS ONE 8(12):e77173. doi:10.1371/journal.pone.0077173

Dykens JA, Shick JM, Benoit C, Buettner GR, Winston GW (1992) Oxygen radical production in the sea-anemone *Anthopleura elegantissima* and its endosymbiotic algae. J Exp Biol 168:219–241

Fabricius KE (2005) Effects of terrestrial runoff on the ecology of corals and coral reefs: review and synthesis. Mar Pollut Bull 50:125–146

Fabricius KE, De'ath G (2004) Identifying ecological change and its causes: a case study on coral reefs. Ecol Appl 14:1448–1465

Fankboner PV (1971) Intracellular digestion of symbiotic zooxanthellae by host amoebocytes in giant clams (Bivalvia: Tridacnidae), with a note on the nutritional role of the hypertrophical siphonal epidermis. Biol Bull 141:222–234

Fitt WK, Cook CB (1990) Some effect of host feeding on growth of zooxanthellae in the marine hydroid *Myrionema amblonense* in the laboratory and in nature. In: Nardon P, Gianinazzi-Pearson V, Grenier AM, Margulis L, Smith DC (eds) Endocytobiology IV. INRA, Paris, pp 281–284

Fitt WK, Warner ME (1995) Bleaching patterns of four species of Caribbean reef corals. Biol Bull 189:298–307

Fitt WK, Brown BE, Warner ME, Dunne RP (2001) Coral bleaching: interpretation of thermal tolerance limits and thermal thresholds in tropical corals. Coral Reefs 20:51–65

Frentzen M (2004) Phosphatidylglycerol and sulfoquinovosyldiacylglycerol: anionic membrane lipids and phosphate regulation. Curr Opin Plant Biol 7:270–276

Fujimura H, Higuchi T, Shiroma K, Arakaki T, Hamdun AM, Nakano Y, Oomori T (2008) Continuous-flow complete-mixing system for assessing the effects of environmental factors on colony-level coral metabolism. J Biochem Biophys Methods 70:865–872

Fujise L, Yamashiya H, Suzuki G, Koike K (2013) Expulsion of zooxanthellae (*Symbiodinium*) from several species of scleractinian corals: comparison under non-stress conditions and thermal stress conditions. Galaxea JCRS 15:29–36

Fukabori Y (1998) Whitening mechanism of reef building coral. Decomposition and discharge of symbiotic algae. Umiushi Tsushin 18:2–4 (in Japanese)

Gates RD (1990) Seawater temperature and sublethal coral bleaching in Jamaica. Coral Reefs 8:193–198

Goericke R, Strom SL, Bell MA (2000) Distribution and sources of cyclic pheophorbides in the marine environment. Limnol Oceanogr 45:200–211

Gomez-Gil B, Soto-Rodríguez S, García-Gasca A, Roque A, Vazquez-Juarez R, Thompson FL, Swings J (2004) Molecular identification of *Vibrio harveyi*-related isolates associated with diseased aquatic organisms. Microbiology (Reading, England) 150:1769–1777

Harris PG, Pearce GES, Peakman TM, Maxwell JR (1995) A widespread and abundant chlorophyll transformation product in aquatic environments. Org Geochem 23:183–187

Higuchi T, Agostini S, Casareto BE, Yoshinaga K, Suzuki T, Nakano Y, Fujimura H, Suzuki Y (2013) Bacterial enhancement of bleaching and physiological impacts on the coral *Montipora digitata*. J Exp Mar Biol Ecol 440:54–60

Hoegh-Guldberg O (1999) Climate change, coral bleaching and the future of the world's coral reefs. Mar Freshw Res 50:839–866

Hoegh-Guldberg O, Smith GJ (1989) The effects of sudden changes in light, temperature and salinity on the population density and export of zooxanthellae from the reef corals *Seriatopora hystrix* and *Stylophora pistillata*. J Exp Mar Biol Ecol 129:279–303

Hoegh-Guldberg O, Mumby PJ, Hooten AJ, Steneck RS, Greenfield P, Gomez E, Harvell CD, Sale PF, Edwards AJ, Caldeira K,

Knowlton N, Eakin CM, Iglesias-Prieto R, Muthiga N, Bradbury RH, Dubi A, Hatziolos ME (2007) Coral reefs under rapid climate change and Ocean acidification. Science 318:1737–1742

Hughes TP, Baird AH, Bellwood DR, Card M, Connolly SR, Folke C, Grosberg R, Hoegh-Guldberg O, Jackson JBC, Kleypas J, Lough JM, Marshall P, Nyström M, Palumbi SR, Pandolfi JM, Rosen B, Roughgarden J (2003) Climate change, human impacts, and the resilience of Coral Reefs. Science 301:929–933

Jones RJ (1997) Changes in zooxanthellar densities and chlorophyll concentrations in corals during and after a bleaching event. Mar Ecol Prog Ser 158:51–59

Karuso P, Bergquist PR, Buckleton JS, Cambie RC, Clark GR, Rickard CEF (1986) 13^2, 17^3-Cyclopheophorbide enol, the first porphyrin isolated from a sponge. Tetrahedron Lett 27:2177–2178

Kashiyama Y, Tamiaki H (2014) Risk management by organisms of the phototoxicity of chlorophylls. Chem Lett 43:148–156

Kashiyama Y, Yokoyama A, Kinoshita Y, Shoji S, Miyashiya H, Shiratori T, Suga H, Ishikawa K, Ishikawa A, Inouye I, Ishida K, Fujinuma D, Aoki K, Kobayashi M, Nomoto S, Mizoguchi T, Tamiaki H (2012) Ubiquity and quantitative significance of detoxification catabolism of chlorophyll associated with protistan herbivory. Proc Natl Acad Sci U S A 109:17328–17335

Kashiyama Y, Yokoyama A, Shiratori T, Inouye I, Kinoshita Y, Mizoguchi T, Tamiaki H (2013) 13^2,17^3-Cyclopheophorbide *b* enol as a catabolite of chlorophyll *b* in phycophagy by protists. FEBS Lett 587:2578–2583

Kashiyama Y, Yokoyama A, Tamiaki H (2014) 13^2,17^3-Cyclopheophorbide enols as detoxified catabolites of chlorophylls – a widely distributed metabolism among phototrophic/heterotrophic protists. Abstract with peer review of 8th international conference on Porphyrins and Phthalocyanines (ICPP) abs121382.

Kleypas JA (1994) A diagnostic model for predicting global coral reef distribution. In: Bellwood O, Choat H, Saxena N (eds) Recent advances in Marine Science and Technology. PACON International and James Cook University, Townsville, pp 211–220

Kuroki T, van Woesik R (1999) Changes in zooxanthellae characteristics in the coral *Stylophora pistillata* during the 1998 bleaching event. Galaxea JCRS 1:97–101

Kushmaro A, Rosenberg E, Fine M, Loya Y (1997) Bleaching of the coral *Oculina patagonica* by Vibrio AK-1. Mar Ecol Prog Ser 147:159–165

Lesser MP, Farrell JH (2004) Exposure to solar radiation increases damage to both host tissues and algal symbionts of corals during thermal stress. Coral Reefs 23:367–377

Lesser MP, Stochaj WR, Tapley DW, Shick JM (1990) Bleaching in coral reef anthozoans: effects of irradiance, ultraviolet radiation, and temperature on the activities of protective enzymes against active oxygen. Coral Reefs 8:225–232

Louda JW, Loitz JW, Rudnick DT, Baker EW (2000) Early diagenetic alteration of chlorophyll-*a* and bacteriochlorophyll-*a* in a contemporaneous marl ecosystem; Florida Bay. Organic Geochem 31:1561–1580

Louda JW, Neto RR, Magalhaes ARM, Schneider VF (2008) Pigment alterations in the brown mussel *Perna perna*. Comp Biochem Physiol B 150:385–394

Marubini F, Davies PS (1996) Nitrate increases zooxanthellae population density and reduces skeletogenesis in corals. Mar Biol 127:319–328

Miller DJ, Yellowlees D (1989) Inorganic nitrogen uptake by symbiotic marine Cnidarians: a critical review. Proc R Soc Lond B Biol Sci 237:109–125

Mise T, Hidaka M (2003) Degradation of zooxanthellae in the coral *Acropora nasuta* during bleaching. Galaxea JCRS 5:33–40

Muller EM, Rogers CS, Spitzack AS, van Woesik R (2007) Bleaching increases likelihood of disease on *Acropora palmata* (Lamarck) in Hawksnest Bay, St John, US Virgin Islands. Coral Reefs 27:191–195

Muscatine L (1990) The role of symbiotic algae in carbon and energy flux in reef corals. In: Coral reefs, ecosystems of the world. Elsevier, Amsterdam, pp 75–87

Muscatine L, Falkowski PG, Porter JW, Dubinsky Z (1984) Fate of photosynthetic fixed carbon in light- and shade-adapted colonies of the symbiotic coral *Stylophora pistillata*. Proc R Soc Lond B 222:181–202

Muscatine L, Falkowski PG, Dubinsky Z, Cook PA, McCloskey LR (1989) The effect of external nutrient resources on the population dynamics of zooxanthellae in a reef coral. Proc R Soc Lond B Biol Sci 236:311–324

Mydlarz LD, Couch CS, Weil E, Smith G, Harvell CD (2009) Immune defenses of healthy, bleached and diseased *Montastraea faveolata* during a natural bleaching event. Dis Aquat Organ. Special Issue 5; The role of environment and microorganisms in diseases of corals. 87: 67–78

Ocampo R, Sachs JP, Repeta DJ (1999) Isolation and structure determination of the unstable 13^2, 17^3-Cyclopheophorbide *a* enol from recent sediments. Geochim Cosmochim Acta 63:3743–3749

Pacher P, Beckman JS, Liaudet L (2007) Nitric oxide and peroxynitrite in health and disease. Physiol Rev 87:315–424

Papina M, Meziane T, van Woesik R (2007) Acclimation effect on fatty acids of the coral *Montipora digitata* and its symbiotic algae. Comp Biochem Physiol B 147:583–589

Parkhill J-P, Maillet G, Cullen JJ (2001) Fluorescence-based maximal quantum yield for PSII as a diagnostic of nutrient stress. J Phycol 37:517–529

Perez S, Weis VM (2006) Cnidarian bleaching and nitric oxide: an eviction notice mediates the breakdown of symbiosis. J Exp Biol 209:2804–2810

Perl-Treves R, Perl A (2002) Oxidative stress: an introduction. In: Van Montagu M, Inzé D (eds) Oxidative stress in plants. Taylor and Francis Books Ltd, London, pp 1–32

Reimer JD, Ono S, Furushima Y, Tsukahara J (2007) Seasonal changes in morphological condition of symbiotic Dinoflagellates (*Symbiodinium* spp.) in *Zoanthus sansibaricus* (Anthozoa: Hexacorallia) in southern Japan. South Pacific Stud 27:1–24

Ritchie K (2006) Regulation of microbial populations by coral surface mucus and mucus-associated bacteria. Mar Ecol Prog Ser 322:1–14

Rosenberg E, Falkovitz L (2004) The *Vibrio shiloi*/*Oculina patagonica* model system of coral bleaching. Annu Rev Microbiol 58:143–159

Rosenberg E, Koren O, Reshef L, Efrony R, Zilber-Rosenberg I (2007) The role of microorganisms in coral health, disease and evolution. Nat Rev Microbiol 5:355–362

Sakata K, Yamamoto K, Ishikawa H, Yagi A, Etoh H, Ina K (1990) Chlorophyllone *a*, a new phaeophorbide *a* related compound isolated from *Ruditapes philippinarum* as an antioxidative compound. Tetrahedron Lett 31:1165–1168

Steele RD, Goreau NI (1977) The breakdown of symbiotic zooxanthellae in the sea anemone *Phyllactis* (= *Oulactis*) *flosculifera* (Actiniaria). J Zool (Lond) 181:421–437

Stimson J, Larned S, Conklin E (2001) Effects of herbivory, nutrient levels, and introduced algae on the distribution and abundance of the invasive macroalga *Dictyosphaeria cavernosa* in Kaneohe Bay, Hawaii. Coral Reefs 19:343–357

Sussman M, Willis BL, Victor S, Bourne DG (2008) Coral pathogens identified for White Syndrome (WS) epizootics in the Indo-Pacific. PloS ONE 3:e2393. doi:10.1371/journal.pone.0002393

Sutherland K, Porter J, Torres C (2004) Disease and immunity in Caribbean and Indo-Pacific zooxanthellate corals. Mar Ecol Prog Ser 266:273–302

Suzuki Y, Casareto BE (2011) The role of Dissolved Organic Nitrogen (DON) in coral biology and reef ecology. In: Dubinsky Z, Stamler N (eds) Coral reefs: an ecosystem in transition. Springer, Dordrecht, pp 207–214

Suzuki T, Casareto BE, Shioi Y, Ishikawa Y, Suzuki Y (2015) Finding of 13^2, 17^3–cyclopheophorbide *a* enol as degradation product of chlorophyll in shrunk zooxanthellae of the coral *Montipora digitata*. J Phycol 51(1):37–45. doi:10.1111/jpy.12253

Szmant AM (2002) Nutrient enrichment on coral reefs: is it a major cause of coral reef decline? Estuaries 25:743–766

Titlyanov EA, Titlyanova TV, Leletkin VA, Tsukahara J, van Woesik R, Yamazato K (1996) Degradation of zooxanthellae and regulation of their density in hermatypic corals. Mar Ecol Prog Ser 139:167–178

Titlyanov EA, Titlyanova TV, Loya Y, Yamazato K (1998) Degradation and proliferation of zooxanthellae in planulae of the hermatypic coral *Stylophora pistillata*. Mar Biol 130:471–477

Titlyanov EA, Titlyanova TV, Yamazato K, van Woesik R (2001) Photo-acclimation of the hermatypic coral *Stylophora pistillata* while subjected to either starvation or food provisioning. J Exp Mar Biol Ecol 257:163–181

Trapido-Rosenthal H, Zielke S, Owen R, Buxton L, Boeing B, Bhagooli R, Archer J (2005) Increased zooxanthellae nitric oxide synthase activity is associated with coral bleaching. Biol Bull 208:3–6

Trench R (1993) Microalgal-invertebrate symbioses – a review. Endocyt Cell Res 9:135–175

Warner ME, Fitt WK, Schmidt GW (1999) Damage to photosystem II in symbiotic dinoflagellates: a determinant of coral bleaching. Proc Natl Acad Sci U S A 96:8007–8012

Watanabe N, Yamamoto K, Ishikawa H, Yagi A, Sakata K, Brinen LS, Clardy J (1993) New chlorophyll-*a*-related compounds isolated as antioxidants from marine bivalves. J Nat Prod 56:305–317

Weis VM (2008) Cellular mechanisms of Cnidarian bleaching: stress causes the collapse of symbiosis. J Exp Biol 211:3059–3066

Wiedenmann J, D'Angelo C, Smith EG, Hunt AN, Legiret FE, Postle AD, Achterberg EP (2012) Nutrient enrichment can increase the susceptibility of reef corals to bleaching. Nat Clim Chang 3:160–164

Yamada N, Tanaka A, Horiguchi T (2013) cPPB-*a*E is discovered from photosynthetic benthic dinoflagellates. J Phycol 50(1):101–107

Yamamoto K, Sakata K, Watanabe N, Yagi A, Brinen LS, Clardy J (1992) Chlorophyllonic acid and methyl ester, a new chlorophyll *a* related compound isolated as an antioxidant from short-necked clam, *ruditapes philippinarum*. Tetrahedron Lett 33:2587–2588

Yamasaki H, Sakihama Y (2000) Simultaneous production of nitric oxide and peroxynitrite by plant nitrate reductase: in vitro evidence for the NR-dependent formation of active nitrogen species. FEBS Lett 468:89–92

Coupling Remote Sensing and Coral Annual Band Data to Investigate the History of Catchment Land Use and Coral Reef Status

3

Hiroya Yamano and Tsuyoshi Watanabe

Abstract

Using a combination of remote sensing (aerial photographic) and geochemical (coral annual band) data, we detected significant increases in sediment and nutrient discharge in response to land development during the past 50 years on Ishigaki Island, Southwest Japan. Correlates of coral reef decline were also detected using remote sensing (aerial photographic) data. Given that historical aerial photographs are available for many areas from as early as the 1900s and that long-lived corals can record paleoenvironmental conditions for more than 100 years, our approach could be applied to reconstructing changes in other coral reef regions worldwide since the early twentieth century or before.

Keywords

Coral reef • Land-based pollution • Remote sensing • Coral annual band

3.1 Introduction

Coral reefs have been in decline in recent decades on account of multiple stressors (e.g., Burke et al. 2011) (Fig. 3.1), including land-based discharge of excess sediment loads, nutrients, and other pollutants (Fabricius 2005; Bartley et al. 2014), which has likely reduced the resilience of corals to other stressors, such as warming temperatures (Hongo and Yamano 2013). Both urban and agricultural developments can increase sediment and nutrient discharge to coastal waters (e.g., Brodie et al. 2012), and the effects of these stressors are therefore especially pronounced on fringing reefs in coastal areas (Burke et al. 2011). Thus, the combined reconstruction of the trajectories of land use and adjacent coral reef status may be a particularly reliable approach for examining the effects of sediment and nutrient discharge on corals and coral reefs.

Aerial photographic data (available for approximately the past 100 years in many areas) (Cochran 2013), satellite data (available since the 1970s), and emerging high-resolution satellite datasets can be used to reconstruct land-use history in coastal catchment areas (Ramos-Scharrón et al. 2015) and the distribution of coral reefs in coastal waters (Yamano 2013). Thus, remote sensing data can provide information about both land-use patterns and coral reef status. However, the acquisition frequency of remote sensing data during the past 100 years has been rather coarse, generally greater than 5 years.

In this study, we used coral skeletal data to overcome the limitation of the coarse temporal resolution of remote sensing data. Massive corals such as *Porites* spp. exhibit annual bands that have been used for paleoenvironmental reconstructions at temporal resolutions of weeks to months (e.g., Gagan et al. 2000). Based on recent advances in analytical technology, it has been shown that variations in the flux rates of land-based pollution into waters inhabited by corals can be estimated from variations in the concentrations of barium (e.g., McCulloch et al. 2003; Sowa et al. 2014) and

H. Yamano (✉)
National Institute for Environmental Studies, 16-2 Onogawa, Tsukuba, Ibaraki 305-8506, Japan
e-mail: hyamano@nies.go.jp

T. Watanabe
Department of Natural History Sciences, Faculty of Science, Hokkaido University, N10W8, Kita-ku, Sapporo, Hokkaido 060-0810, Japan

H. Kayanne (ed.), *Coral Reef Science*, Coral Reefs of the World 5, DOI 10.1007/978-4-431-54364-0_3

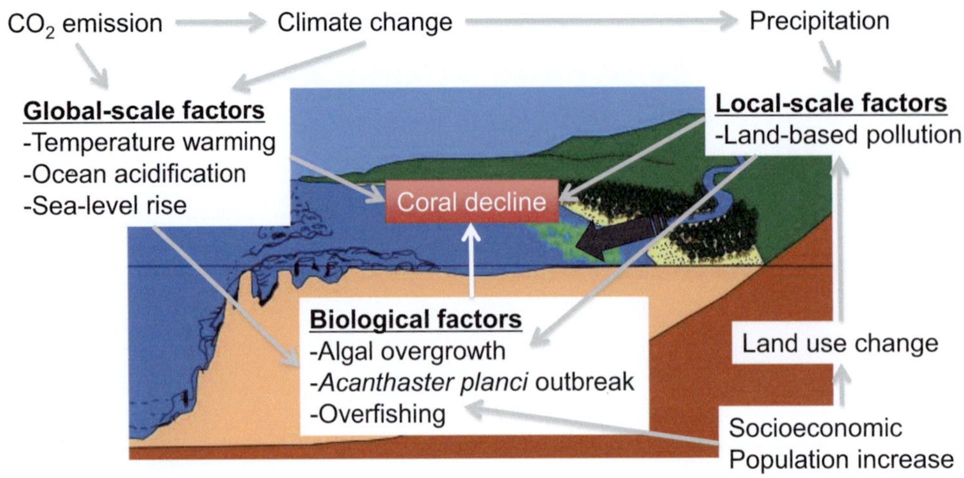

Fig. 3.1 Multiple stressors to coral reefs

heavy metals (e.g., Lewis et al. 2007) in coral skeletons and, similarly, that $\delta^{15}N$ values in corals are a good proxy for nutrient fluxes (e.g., Marion et al. 2005; Yamazaki et al. 2011).

This chapter provides an overview of our work to couple remote sensing and coral annual band data in order to investigate the history of land development and coral reef decline on Ishigaki Island, Southwest Japan. Remote sensing data reveal decadal-scale changes in land-use patterns and coral/seagrass distributions, and coral skeletons record variability in heavy metal concentrations and isotopic fractionation of nutrient-derived N that appear to have been influenced by land-use development in coastal catchment areas.

3.2 Field Setting

In the Ryukyu Islands, Southwest Japan, extensive urban and agricultural development after the 1953 reversion of the Amami Archipelago (south of Kagoshima Prefecture) and the 1972 reversion of Okinawa Prefecture to Japan resulted in a significant increase in sediment and nutrient discharge to coastal waters (Omija 2004), resulting in severe decline in the health of both river and coastal ecosystems (e.g., West and van Woesik 2001; Omija 2004). In response to environmental degradation related to sediment discharge, Okinawa Prefecture established the Okinawa Prefecture Red Soil Erosion Prevention Ordinance in 1994, with enforcement initiated in October 1995. As a result, sediment discharge from construction sites was successfully restricted; however, sediment discharge from farmlands is not yet fully regulated, and present-day sugarcane farms continue to be a significant source of sediment pollution.

We studied a coastal area on Ishigaki Island, located south of the Ryukyu Islands, Japan, where the Todoroki River supplies large amounts of sediment and nutrients to coastal reefs. The river flows through 15 km^2 of sugarcane, pasture, pineapple, and paddy fields distributed across the drainage basin (Nakasone et al. 2001; Paringit and Nadaoka 2003; Hasegawa 2011). The average annual precipitation in the catchment area is ~2000 mm, 60 % of which falls during the rainy season in May and June or during typhoons in August and September; these periods are also the periods of heaviest soil erosion and potentially excessive discharge of sediment and nutrients to coast waters. Paringit and Nadaoka (2003) developed a theoretical model for sediment discharge from farmlands in this area. Farmlands are also suggested to be the source of high concentrations of nutrients in river water and groundwater (Blanco et al. 2008, 2010, 2011), thus contributing to the pollution of coastal waters.

Shiraho Reef, located on the east coast of Ishigaki Island, Southwest Japan, facing the Pacific Ocean (Fig. 3.2), is situated in a monsoon area where the dominant winds are from the south (southerlies) in summer and from the north (northerlies) in winter (Yamano et al. 1998; Hasegawa and Yamano 2004). Over the year, northerly winds are dominant, and the speeds of the northerly winds are generally greater than those of the southerly winds. Ocean currents off the mouth of the Todoroki River generally flow northward, but southward flows occur during high swells and/or northerly winds (Yamano et al. 2014). Under southward flow conditions, sediment and nutrients discharged from the Todoroki River could be transported onto Shiraho Reef, thereby affecting corals on the reef.

Figure 3.2 shows the geographic setting of Shiraho Reef, the well-developed fringing reef structure, and the distinct topographic zonation from land to ocean which includes a back-reef moat and a reef crest. Shiraho Reef extends for ~3 km along the shore, and the distance from the shoreline to the reef crest is 700–850 m. At the northern end of the Shiraho Reef, a prominent channel connects the back-reef moat to the open ocean (Nadaoka et al. 2001; Tamura

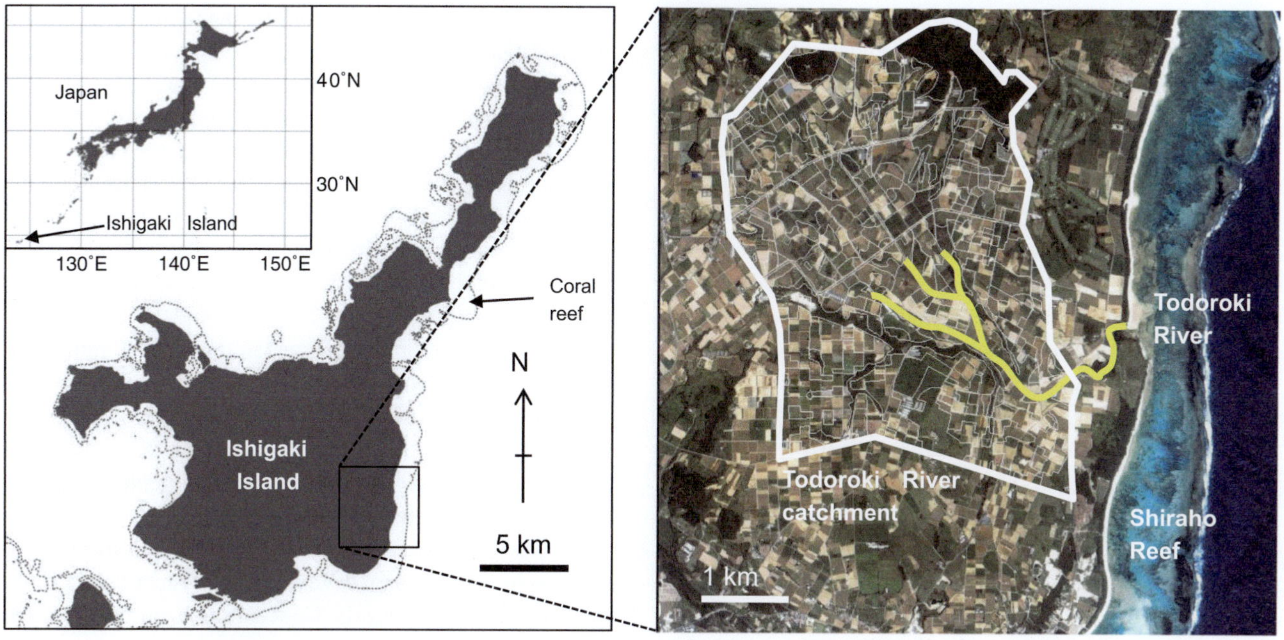

Fig. 3.2 Location of Ishigaki Island, the Todoroki River catchment area, and Shiraho Reef

et al. 2007). In the central and southern portions of the reef, small shallow (1–2 m deep) channels transport water from the back-reef moat to the ocean. At the southern boundary of the reef, a shallow area called "Watanji" separates the back-reef moat from the ocean.

As on other reef systems, distinct ecological zonation is found on Shiraho Reef (Nakamori et al. 1992; Iryu et al. 1995; Kayanne et al. 2002). Five zones are observed from shore to ocean: (1) seagrass beds with small patches of *Porites* corals close to the beach; (2) sand bottom in the back-reef moat; (3) branching *Acropora* and *Montipora* corals at the back of the reef crest; (4) larger benthic foraminifera on the reef crest, which experiences subaerial exposure during low tides; and (5) robust-branching *Acropora* corals at the reef edge facing the ocean. In the southern part of the reef, the blue coral *Heliopora coerulea*, instead of *Acropora* and *Montipora* corals, occurs in zone (3).

3.3 Reconstructing Land Use and River Discharge

3.3.1 Remote Sensing

Land-use changes in the Todoroki River catchment area were analyzed using aerial photographs and satellite image data. Classification categories were as follows: forest (evergreen), grassland, sugarcane, pineapple, and paddy. Aerial photographs taken in 1962, 1972, 1977/1978, 1986, 1989, 1991, and 1995 were analyzed by visual inspection and interpretation (Hasegawa 2011). Then, a recent land-use map was generated from satellite data by considering the crop calendar that shows dynamics of growth patterns of the categories (Ishihara et al. 2014). Six satellite images, which included images from winter and summer, for the period 2006–2008, were used to derive a normalized difference vegetation index (NDVI), calculated using the radiance values from all six images as follows:

$$\mathrm{NDVI} = (R_{\mathrm{NIR}} - R_{\mathrm{Red}})/(R_{\mathrm{NIR}} + R_{\mathrm{Red}}) \quad (3.1)$$

where R_{Red} and R_{NIR} are the red and near-infrared radiance values, respectively. Thus, growth patterns could be detected by NDVI in the six satellite images. Then, a decision tree method, the structure of which was designed according to the crop calendar, was applied to classify the land-use type into one of the five categories described above.

3.3.2 Coral Annual Bands

Coral cores were collected from sub-tidal areas of the back reef at depths of a few meters below low tide. The cores were collected using a handheld air drill driven by scuba tanks (Adachi and Abe 2003). Precast concrete plugs were inserted to prevent colonization of the inside of the corals by boring organisms and to allow coral overgrowth. Cores were drilled vertically and horizontally to obtain the clearest growth banding pattern along the maximum growth axis.

After collection, cores were rinsed in fresh water and air-dried. Slices (~2–5 mm thick) were cut along axes of the main growth using a rock saw equipped with a diamond-tipped blade cooled with water; the slices were then uniformly planed. Annual bands were revealed by X-ray images. Because of the possible existence of pseudo-annual bands, an age model for the cores was calculated by matching variations in U concentrations, as U concentrations are well correlated with sea surface temperature (SST) variations and were measured at the same time as Fe and Mn concentrations on the same subsamples (Min et al. 1995). Preliminary high-resolution $\delta^{18}O$ data (1-mm subsampling interval) were also used in the construction of the age model. After age determination, heavy metals (Mn and Fe) and nitrogen isotopic ratios ($\delta^{15}N$) of the skeletons were analyzed (Inoue et al 2014; Yamazaki et al 2014, respectively) at 3-mm subsampling intervals, corresponding to time intervals of ca. 2 months. Pretreatment of the cores, including slicing, X-ray imaging, and subsampling, was performed at the Coral Core Center, Hokkaido University, Japan (http://ccc.sci.hokudai.ac.jp/Coral_Core_Center/Welcome.html). Detailed analytical procedures for determinations of Mn and Fe are provided by Inoue et al. (2014) and for $\delta^{15}N$ are provided by Yamazaki et al. (2014).

3.4 Reconstructing Coral Reef Status

Detailed distributions of seagrass and coral were mapped using aerial photographs (Hasegawa 2011; Harii et al. 2014). The visual interpretation of seagrass areas was based on texture, using the same dataset as that used to reconstruct land-use patterns. Seagrass areas are confined to sandy bottoms close to shorelines (Hasegawa 2011) and are easy to identify on both color and grayscale photographs. Areas of high coral cover (50–100 %) and medium coral cover (5–50 %) were detected according to the supervised classification of color aerial photographs taken in 1995, 2000, 2004, and 2006, based on in situ transect survey information (Harii et al. 2014). Contextual editing was performed to correct erroneous classifications along the shoreline and reef crest where corals do not exist, and areas of each category were recalculated.

3.5 Coupling of Land Use and Coral Reef Status

Historical patterns of land use in the Todoroki River catchment area, concentrations of heavy metals and nitrogen isotopes in *Porites* corals, changes in the area of seagrass and coral cover on Shiraho Reef, and related patterns of land development (i.e., the societal background) are arranged in chronological order in Fig. 3.3. Land-use patterns show significant changes during this time, including rapid expansion of sugarcane fields after the 1970s (Fig. 3.3b) and extensive agricultural development starting in 1972 (Fig. 3.3a). Before 1980, paddy fields were one of the main uses of land in the catchment area; however, the paddy fields were largely replaced by sugarcane fields starting in the 1980s. Moreover, after the sugarcane harvest in January–March, the sugarcane fields are barren ground before August–September, if the farmers would choose summer planting. Thus, the sugarcane fields could be a source of sediment observed in runoff water during the rainy and typhoon seasons. Paddy fields, on the other hand, may have acted as sediment reservoirs, thereby preventing sediment discharge. The straightening of the Todoroki River between 1978 and 1985 (Fig. 3.3a) also could have enhanced sediment and nutrient transport to the sea.

The change in sediment discharge estimated from aerial photographs and human development is recorded in coral annual bands (Fig. 3.3c, d). Concentration profiles of both Mn and Fe show significant increases in these metals starting in the 1980s (Inoue et al. 2014) (Fig. 3.3c). Detailed examination of the profiles indicates that 1975 was a turning point with regard to annual variations in the mean concentrations of the two elements. Before 1975, variations in Fe and Mn were not correlated with one another, whereas during 1976–2004, annual variations of the two elements were synchronous. These patterns suggest that variations in the two elements during 1976–2004 were caused by the same factor(s), i.e., increased sediment loading due to land development.

The variations in $\delta^{15}N$ were not similar to those of the Fe and Mn proxies, suggesting that terrestrial nitrate discharge is not simultaneous with sediment loading (Fig. 3.3c, d). Nutrient discharge, however, could have changed after the 1980s as discussed by Yamazaki et al. (2014) as follows. After the 1980s, the nitrate concentration in the river water was three times as high as the nitrate concentration before the 1980s, assuming that (1) manure fertilizer was used in the paddy fields before the 1980s (Fig. 3.3a), (2) the rate and magnitude of nitrate transport into the coral reef due to the precipitation rate before and after 1980 were the same, and (3) current values of nitrate in the sugarcane soil and manure were applied to the calculation.

The increase in sediment and nutrient discharge from the catchment area appears to be associated with an increase in the area of seagrass coverage and a decrease in the area of coral coverage (Fig. 3.3e, f). Based on analyses of aerial photographs, Hasegawa (2011) reported that seagrass coverage on Shiraho Reef increased from 8.6 ha in 1972 to 55.0 ha in 2004 (Fig. 3.3e), while Harii et al. (2014) reported that the total area of high and medium coral cover decreased from 2.8 ha in 1995 to 0.8 ha in 2006 (Fig. 3.3f). The main causes of decreased coral cover have been regarded as bleaching

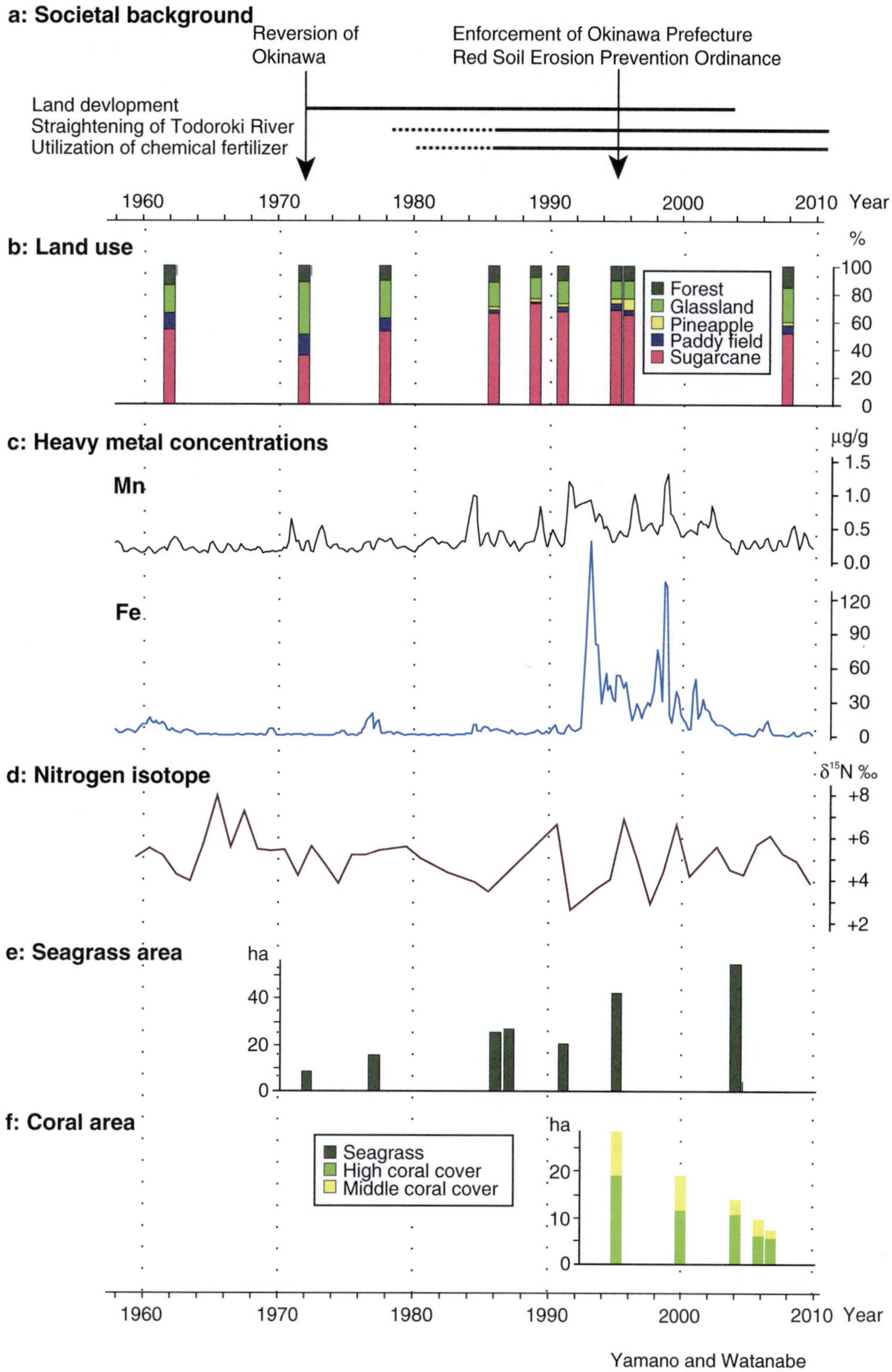

Fig. 3.3 Figure integrating data from (**a**) societal factors, (**b**) historic changes in land use in the Todoroki River catchment area, (**c**) heavy metal concentrations and (**d**) nitrogen isotopes for *Porites* corals, and (**e**) areas of seagrass and (**f**) areas of coral on Shiraho Reef (**e**, **f**). References: (**a**) Hasegawa (2011) and Yamazaki et al. (2014), (**b**) Hasegawa (2011), (**c**) Inoue et al. (2014), (**d**) Yamazaki et al. (2014), (**e**) Hasegawa (2011), and (**f**) Harii et al. (2014)

events and typhoons, although land-based pollution may have had additional chronic negative effects on corals, as suggested by Hongo and Yamano (2013). The increased presence of blue-green phytoplankton and benthic blue-green microalgae on Shiraho Reef also suggests the impact of increased nutrient loadings (Blanco et al. 2008).

Collectively, these results show a clear association between processes in the catchment area and those on the reef: extensive land development and an increase in sugarcane cultivation resulted in increased sediment and nutrient discharge, which was recorded in coral skeletons as changes in Fe and Mn concentrations and $\delta^{15}N$ values, and which resulted in an increase in the area of seagrass and a decrease in the area of coral.

3.6 Discussion and Conclusions

In this study, we presented a novel procedure that couples trajectories of land use, sediment and nutrient discharge, and coral reef status at a decadal time scale and at the spatial scale of a catchment area and an adjacent coral reef. Further investigations applying process-based experimental results would enhance the robustness of the results. Because historical aerial photographs may be available from as early as the 1900s (Cochran 2013), examination of land-use changes using aerial photographs, along with recent high-resolution satellite images, as well as a review of historical literature and socioeconomic data, can aid in historical reconstructions of land-use development and societal contexts. Coral annual bands have enormous potential for reconstructing the dynamics of sediment and nutrient discharge over periods of 50 years or more, and even more long-lived coral samples are available. Thus, our approach could be applied to reconstructing changes in coral reef regions worldwide during the twentieth century.

The coupling of remote sensing and coral annual band data can improve knowledge of the historical characteristics of land use and may also help to establish guidelines for land management policies to reduce pollutant discharge. Of interest in our particular case is the potential for a decrease in sediment discharge after 2004, as detected by Mn and Fe profiles (Fig. 3.3), which could be an outcome of specific conservation policies, especially the Okinawa Prefecture Red Soil Erosion Prevention Ordinance and the setting of green belts around sugarcane fields by the World Wildlife Fund (WWF) of Japan. The effects of such actions could be examined in more in detail in future work.

Corals not only record local land-based pollution but also global climate change (Gagan et al. 2000). This is true of corals on Shiraho Reef, which record a shift in climatic regime from East Asian monsoon to El Nino/Southern Oscillation (Tsunoda et al. 2008; Mishima et al. 2010). Further advances in analytical technology may allow reconstructions of pH changes related to ocean acidification (Pelejero et al. 2005; Shinjo et al. 2013), which is another climate change issue related to CO_2 emissions. Therefore, coral annual band analyses may allow for more integrated analyses of global and local environmental changes (Fig. 3.1). Harii et al. (2014) provide a good example of the need for more integrated models, in their consideration of multiple disturbances that affect the dynamics of coral reefs and coral reef status.

In addition, both recent and fossil coral records provide information on environmental conditions. Sowa et al. (2014) analyzed fossil *Porites* dated at 3000 and 1000 years old and suggested that seasonality of Ba/Ca ratios may provide yet another proxy of sediment discharge (McCulloch et al. 2003) in prehistoric times. Elemental Ba/Ca ratios may be related to agricultural activities, and this possibility could be examined in collaboration with archaeological investigations.

A consideration of catchment-to-reef continua is important for establishing proper land-based management plans for the conservation of coral reefs. Excellent outcomes related to this issue are emerging (e.g., as reported in special issues of the *Marine Pollution Bulletin* and *Agriculture, Ecosystems and Environment*; Delvin and Schaffelke 2012; Thorburn 2013). Our work similarly contributes to the future of effective land management practices.

Acknowledgments This work was supported by a Grant-in-Aid for Scientific Research on Innovative Areas "Coral reef science for symbiosis and coexistence of humans and ecosystems under combined stresses" (No. 20121004) of the Ministry of Education, Culture, Sports, Science and Technology (MEXT), Japan. We thank collaborators in team B01 "Historic changes of coral reefs and their stressors," who have published informative and insightful results. We also thank the residents of Shiraho village and WWF Japan for providing this research opportunity.

References

Adachi H, Abe O (2003) "Air drill" for submerged coral drilling. Mar Technol Soc J 37:31–36

Bartley R, Bainbridge ZT, Lewis SE, Kroon FJ, Wilkinson SN, Brodie JE, Silburn DM (2014) Relating sediment impacts on coral reefs to watershed sources, processes and management: a review. Sci Total Environ 468–469:1138–1153

Blanco AC, Nadaoka K, Yamamoto T (2008) Planktonic and benthic microalgal community composition as indicators of terrestrial influence on a fringing reef in Ishigaki Island, Southwest Japan. Mar Environ Res 66:520–535

Blanco AC, Nadaoka K, Yamamoto T, Kinjo K (2010) Dynamic evolution of nutrient discharge under stormflow and baseflow conditions in a coastal agricultural catchment in Ishigaki Island, Okinawa, Japan. Hydrol Process 24:2601–2616

Blanco AC, Watanabe A, Nadaoka K, Motooka S, Herrera EC, Yamamoto T (2011) Estimation of nearshore groundwater discharge and its potential effects on a fringing coral reef. Mar Pollut Bull 62:770–785

Brodie JE, Kroon FJ, Schaffelke B, Wolanski EC, Lewis SE, Devlin MJ, Bohnet IC, Brainbridge ZT, Waterhouse J, Davis AM (2012) Terrestrial pollutant runoff to the Great Barrier Reef: an update of issues, priorities and management responses. Mar Pollut Bull 65:81–100

Burke L, Reytar K, Spalding M, Perry A (2011) Reefs at risk revisited. World Resources Institute, Washington, DC

Cochran SA (2013) Photography applications. In: Goodman JA, Purkis SJ, Phinn S (eds) Coral reef remote sensing. Springer, Dordrecht, pp 29–49

Delvin M, Schaffelke B (2012) Catchment-to-reef continuum: case studies from the Great Barrier Reef. A special issue – Marine Pollution Bulletin 2012. Mar Pollut Bull 65:77–80

Fabricius KE (2005) Effect of terrestrial runoff on the ecology of corals and coral reefs: review and synthesis. Mar Pollut Bull 50:125–146

Gagan MK, Ayliffe LK, Beck JW, Cole JE, Druffel ERM, Dunbar RB, Schrag DP (2000) New views of tropical paleoclimate from corals. Quat Sci Rev 19:45–64

Harii S, Hongo C, Ishihara M, Ide Y, Kayanne H (2014) Impacts of multiple disturbances on coral communities at Ishigaki Island, Okinawa, Japan, during a 15 year survey. Mar Ecol Prog Ser 509:171–180

Hasegawa H (2011) The decline of coral reef conditions caused by the extensive land modification: a case study of the Shiraho area on Ishigaki Island, Okinawa, Japan. J Remote Sens Soc Jpn 31:73–86

Hasegawa H, Yamano H (2004) Ishigaki Island. In: Ministry of the Environment and Japanese Coral Reef Society (ed) Coral reefs of Japan. Ministry of the Environment, Tokyo, pp 212–218

Hongo C, Yamano H (2013) Species-specific responses of corals to bleaching events on anthropogenically turbid reefs on Okinawa Island, Japan, over a 15-year period (1995–2009). PLoS ONE 8:e60952

Inoue M, Ishikawa D, Miyaji T, Yamazaki A, Suzuki A, Yamano H, Kawahata H, Watanabe T (2014) Evaluation of Mn and Fe in coral skeletons (*Porites* spp.) as proxies for sediment loading and reconstruction of 50 yrs of land use on Ishigaki Island, Japan. Coral Reefs 33:363–373

Iryu Y, Nakamori T, Matsuda S, Abe O (1995) Distribution of marine benthic organisms and its geological significance in the modern reef complex of the Ryukyu Islands. Sediment Geol 99:243–258

Ishihara M, Hasegawa H, Hayashi S, Yamano H (2014) Land cover classification using multi-temporal satellite images in a subtropical area. In: Nakano S, Yahara T, Nakashizuka T (eds) The biodiversity observation network in the Asia-Pacific region: integrative observations and assessments of Asian biodiversity. Springer, Tokyo, pp 231–237

Kayanne H, Harii S, Ide Y, Akimoto F (2002) Recovery of coral reef populations after the 1998 bleaching on Shiraho Reef in the southern Ryukyus, NW Pacific. Mar Ecol Prog Ser 239:93–103

Lewis SE, Shields GA, Kamber BS, Lough JM (2007) A multi-trace element coral record of land-use changes in the Burdekin River catchment, NE Australia. Palaeogeogr Palaeoclimatol Palaeoecol 246:471–487

Marion GS, Dunbar RB, Mucciarone DA, Kremer JN, Lansing JS, Arthawiguna A (2005) Coral skeletal $\delta^{15}N$ reveals isotopic traces of an agricultural revolution. Mar Pollut Bull 50:931–944

McCulloch M, Fallon S, Wyndham T, Hendy E, Lough J, Barnes D (2003) Coral record of increased sediment flux to the inner Great Barrier Reef since European settlement. Nature 421:727–730

Min GR, Edwards RL, Taylor FW, Récy J, Gallup CD, Beck JW (1995) Annual cycles of U/Ca in coral skeletons and U/Ca thermometry. Geochim Cosmochim Acta 59:2025–2042

Mishima M, Suzuki A, Magao M, Ishimura T, Inoue M, Kawahara H (2010) Abrupt shift toward cooler condition in the earliest 20th century detected in a 165 year coral record from Ishigaki Island, southwestern Japan. Geophys Res Lett 37:L15609

Nadaoka K, Nihei Y, Kumano R, Yokobori T, Omija T, Wakaki K (2001) A field observation on hydrodynamic and thermal environments of a fringing reef at Ishigaki Island under typhoon and normal atmospheric conditions. Coral Reefs 20:387–398

Nakamori T, Suzuki A, Iryu Y (1992) Water circulation and carbon flux on Shiraho coral reef of the Ryukyu Islands, Japan. Cont Shelf Res 12–7(8):951–970

Nakasone K, Higa E, Omija T, Yasumura S, Nadaoka K (2001) Measurements of suspended solids in Todoroki River, Ishigaki Island. Ann Rep Okinawa Prefectural Inst Health Environ 35:93–102

Omija T (2004) Terrestrial inflow of soils and nutrients. In: Ministry of the Environment and Japanese Coral Reef Society (ed) Coral reefs of Japan. Ministry of the Environment, Tokyo, pp 64–68

Paringit EC, Nadaoka K (2003) Sediment yield modelling for small agricultural catchments: land-cover parameterization based on remote sensing data analysis. Hydrol Process 17:1845–1866

Pelejero C, Calvo E, McCulloch MT, Marshall JF, Gagan MK, Lough JM, Opdyke BN (2005) Preindustrial to modern interdecadal variability in coral reef pH. Science 309:2204–2207

Ramos-Scharrón CE, Trres-Pulliza D, Hernández-Delgado EA (2015) Catchment- and island wide-scale land cover changes in Puerto Rico (1930s–2004) and their potential effects on coral reef ecosystems. Sci Total Environ 506–507:241–251

Shinjo R, Asami R, Huang K-F, You C-F, Iryu Y (2013) Ocean acidification trend in the tropical North Pacific since the mid-20th century reconstructed from a coral archive. Mar Geol 342:58–64

Sowa K, Watanabe T, Kan H, Yamano H (2014) Influence of land development on the calcification of Holocene *Porites* coral at Nagura Bay, Ishigaki Island, Japan. PLoS One 9:e88790

Tamura H, Nadaoka K, Paringit EC (2007) Hydrodynamic characteristics of a fringing coral reef on the east coast of Ishigaki Island, southwest Japan. Coral Reefs 26:17–34

Thorburn P (2013) Catchments to reef continuum: minimising impacts of agriculture on the Great Barrier Reef. Agric Ecosyst Environ 180:1–3

Tsunoda T, Kawahata H, Suzuki A, Minoshima K, Shikazono N (2008) East Asian monsoon to El Niño/Southern Oscillation: a shift in the winter climate of Ishigaki Island accompanying the 1988/1989 regime shift, based on instrumental and coral records. Geophys Res Lett 35:L13708

West K, van Voesik R (2001) Spatial and temporal variance of river discharge on Okinawa (Japan): inferring the temporal impact on adjacent coral reefs. Mar Pollut Bull 42:864–872

Yamano H (2013) Multispectral applications. In: Goodman JA, Purkis SJ, Phinn SR (eds) Coral reef remote sensing. Springer, Dordreht, pp 51–78

Yamano H, Kayanne H, Yonekura N, Nakamura H, Kudo K (1998) Water circulation in a fringing reef located in a monsoon area: Kabira Reef, Ishigaki Island, Southwest Japan. Coral Reefs 17:89–99

Yamano H, Hata H, Miyajima T, Nozaki K, Kato K, Negishi A, Tamura M, Kayanne H (2014) Water circulation in a fringing reef and implications for coral distribution and resilience. In: Nakano S, Yahara T, Nakashizuka T (eds) The biodiversity observation network in the Asia-Pacific region: integrative observations and assessments of Asian biodiversity. Springer, Tokyo, pp 275–293

Yamazaki A, Watanabe T, Tsunogai U (2011) Nitrogen isotopes of organic nitrogen in reef coral skeletons as a proxy of tropical nutrient dynamics. Geophys Res Lett 38:L19605

Yamazaki A, Watanabe T, Tsunogai U, Hasegawa H, Yamano H (2014) The coral $\delta^{15}N$ record of terrestrial nitrate loading varies with river catchment land use. Coral Reefs 34:353–362

A Review of Coral Studies of the Ryukyu Island Arc to Reconstruct Its Long-Term Landscape History

4

Toru Yamaguchi

Abstract

Coral studies of the Ryukyu Island Arc have resulted in the accumulation of a vast knowledge quite relevant to a future reconstruction or interpretation of its landscape history, i.e., the process of viewing a present landscape as a historical artifact contingently and cumulatively changed through interactions between two types of agents, natural and human. For the geohistory of the late Miocene to the late Holocene, in particular, I review three topics: insularization of the Ryukyu Arc, coral growth during the last glacial maximum, and the formation of a coral reef environment in the middle Holocene. The latter two topics are considered in connection with archaeological and anthropological evidence of late Pleistocene and Holocene human settlement and subsistence. The latest phase after the middle Holocene, however, appears to still not be fully examined in geohistorical coral studies; thus, I briefly describe my geoarchaeological surveys of the shallow sea and alluvial land in the Nagura region of Ishigaki Island, the future examination of which will make us aware of the difficulty of simply determining anthropogenic changes as either environmental degradation or landscape enhancement.

Keywords

Coral studies • Landscape history • Geoarchaeology • Ryukyu Island Arc

4.1 Introduction

A present landscape can be viewed as a historical artifact of long-term interactions between two types of agents, natural and human. The holistic elucidation of such a landscape should be facilitated by transdisciplinary approaches that bridge the natural sciences and the humanities. For instance, archaeology in the Pacific through the past 20 years has contributed substantially to understanding this issue from the viewpoint of historical ecology, which incorporates data collected from a variety of natural sciences, including geology, geomorphology, paleontology, and paleobotany, as noted by Kirch and Hunt (1997) in their landmark volume *Historical Ecology in the Pacific Islands*. It is noteworthy that many studies have used the term "landscape" to scrutinize long-term processes of diachronic interactions between natural processes and anthropogenic actions, probably because positioning the notion of landscape between the two enables the smooth combination of the natural sciences and humanities. Following these studies (see also Yamaguchi et al. 2009), I apply the term "landscape history" to a transdisciplinary approach for examining the diachronic processes that have formed specific present landscapes, and I review coral studies of the Ryukyu Island Arc, one of the well-studied fields in Japan, through a variety of natural sciences, making the work relevant to human settlement history. I also provide a short description of the preliminary outcomes of a geoarchaeological survey of Ishigaki Island in the Yaeyama Islands.

T. Yamaguchi (✉)
Department of Ethnology and Archaeology, Keio University, Tokyo, Mita, 2-15-45, Japan
e-mail: toru38@flet.keio.ac.jp

H. Kayanne (ed.), *Coral Reef Science*, Coral Reefs of the World 5, DOI 10.1007/978-4-431-54364-0_4

4.2 Long-Term Environmental History of Insularization of the Ryukyu Arc

From Tanegashima Island, the northeasternmost island, to Yonaguni-jima Island, the southwesternmost island, many islands are scattered along the Ryukyu Arc, extending for more than 1100 km on the narrow sea ridge from east of the Okinawa Trough (back-arc basin) to west of the Ryukyu Trench, along the boundary between the subducting Philippine Plate and overriding Eurasian Plate. The present island arc is geologically and geographically divided into three regions, the northern, central, and southern Ryukyu, by the Tokara and Kerama (or Miyako) Gaps, each measuring approximately 1000 m deep. The southern Ryukyu is also separated from Taiwan by the 800-m deep Yonaguni Depression (Fig. 4.1a).

Massive silty and sandy sediments of the Shimajiri Group, which appear to have been deposited since the late Miocene, extensively lie on various basement rocks. As there is substantially less earlier marine sediment dating back to the middle Miocene, the former Ryukyu Arc probably formed a part of the eastern margin of the Eurasian Continent (Kizaki 1980; Machida et al. 2001:301–2). New Quaternary nannofossil biostratigraphic data date the deposition of the marine siltstone to 1.552 ± 0.154 Ma, so the arc would have since been separated from the continent (Osozawa et al. 2012). Around 1 Ma, in the early Pleistocene, the succeeding marine environment surrounding the Ryukyu Arc produced a remarkable change in the sediments, from silty and sandy deposits to the carbonate rocks of the Ryukyu Group formed in the Ryukyu Coral Sea.

Two hypotheses have thus far been proposed to explain the main factors promoting the coral growth at this high latitude; one is the appearance of the Okinawa Trough and its trapping of muddy fluvial sediments from East China and Taiwan just before the Ryukyu Arc (Nakagawa 1983), and the other is the entrance of the Kuroshio (Black Current) to the Okinawa Trough and its transport of a tropical warm water mass to the Ryukyu Arc (Koba 1992). The Kuroshio could have entered the Okinawa Trough through the

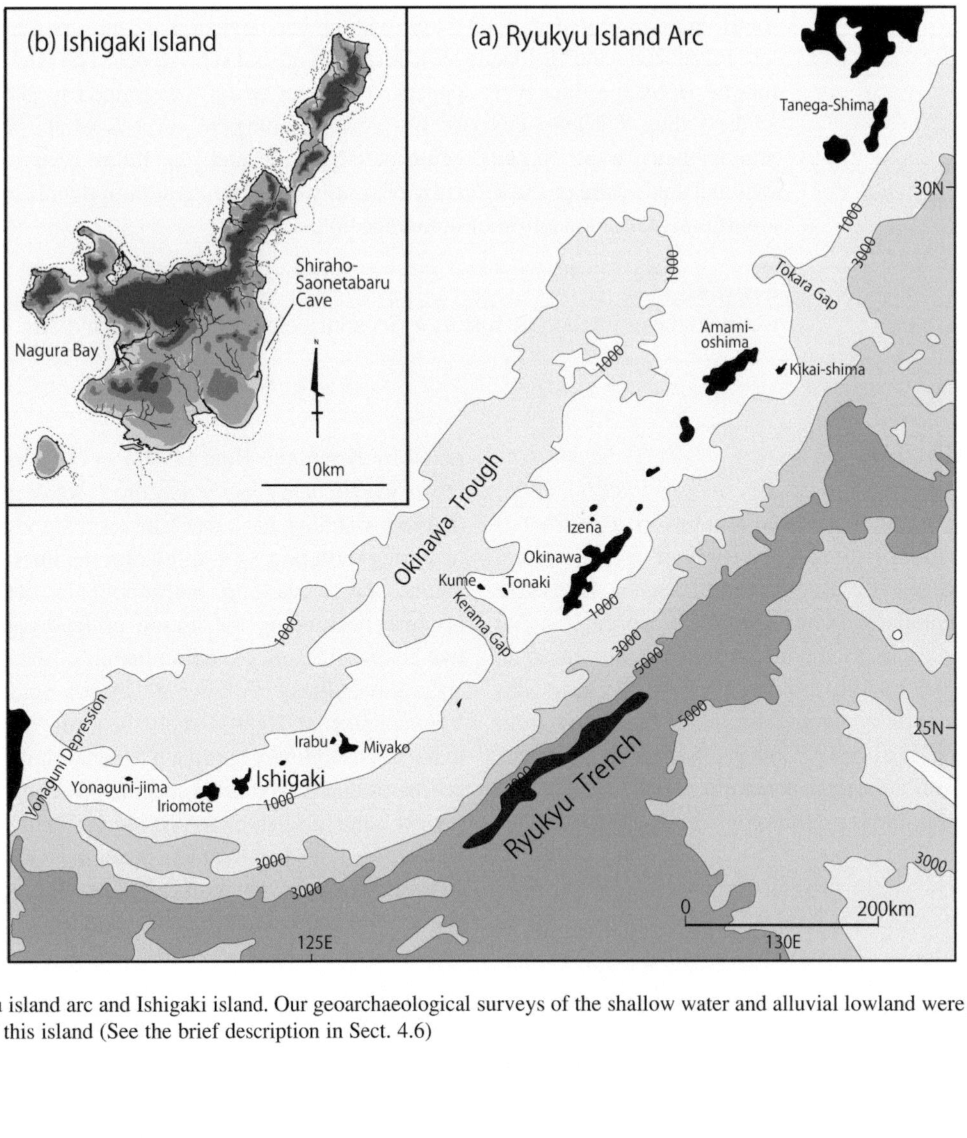

Fig. 4.1 Ryukyu island arc and Ishigaki island. Our geoarchaeological surveys of the shallow water and alluvial lowland were performed in the Nagura region of this island (See the brief description in Sect. 4.6)

deepening and widening of the Yonaguni Depression that would have occurred as the result of the collision of the Luzon Volcanic Arc with Taiwan before 0.7–0.6 Ma (Koba 1992). This deepening and widening of the depression also separated the China–Okinawa land bridge and promoted insularity of the Ryukyu Arc.

Located at a plate boundary, the arc appears to have been subjected to highly complex vertical and horizontal movements that occurred along with glacioeustatic changes in sea level. Therefore, various paleogeographical hypotheses on land configuration changes in the Pleistocene have been proposed, and paleobiological evidence of fossilized terrestrial vertebrates and their associated, but less reliable, radiometric ages have also confused the issue. However, the recent advance of phylogeographical and molecular phylogenetic studies of terrestrial flightless vertebrates has provided a different set of convincing evidence on the historical process of insularization, or geographic isolation, of the islands (Hikida and Ota 1997; Ota 1998, 2003; Matsui et al. 2005). Although the details of this hypothesis are still somewhat inconsistent with newly examined Quaternary nannofossil biostratigraphic data from Shimajiri Group siltstone and overlying sediments (Osozawa et al. 2012), the Ryukyu Arc likely became an insular environment from the formation of the Ryukyu Group, beginning at latest by the early Pleistocene.

4.3 An Insular Environment with Coral Growth Even During the Last Glacial Maximum

Remarkable progress in calcareous nannofossil and sedimentological studies of the Ryukyu Group carbonates by the COral-REef Front (COREF) Project (Matsuda et al. 2012) has revealed that the group includes rhodolith and large foraminifera limestones in addition to reefal sediments deposited intermittently during the early and middle Pleistocene. The aggradational part indicates two long-term trends of Pleistocene sea-level change: the rhodolith and large foraminifera limestones would have formed on island shelf in deeper sea during periods of sea-level highstand, and the latter reefal sediments would have been deposited in shallow water during periods of sea-level lowstand under the condition of continuous tectonic subsidence (Ikeda et al. 1991; Obata and Tsuji 1992; Sagawa et al. 2001).

Obata and Tsuji (1992) analyzed high-resolution seismic reflection profiles and depositional facies distributed from the insular shelf to the shelf slope of the western offshore regions of Miyako and Irabu Islands and summarized the geohistory of the southern Ryukyu. They suggested that tectonic movement culminated with the upheaval of the Ryukyu Arc and that the long-term tectonic trend shifted at 0.4 Ma from subsidence to uplift. This shift also caused a part of the Ryukyu Limestone deposited between ca. 1.2 Ma and 0.4 Ma to become emergent on land.

Recent surveys of the insular shelf have revealed mound structures, characterized by strong reflections and chaotic internal structure in high-resolution seismic reflection profiles, in both central and southern Ryukyu. One of these, at the Amami Spur, is delineated by 80–120-m isobaths northeast of Amami-Oshima Island in the northern part of the central Ryukyu and had a measured thickness of 15 m and width of 400 m. Some irregularly shaped topographic highs were also detected at the shelf edge southwest of Kikai-shima at the northern edge of the central Ryukyu (Matsuda et al. 2011). These structures are considered biogenic reefs or banks consisting of coarse-grained bioclasts deposited during a stage of sea-level lowstand. High-resolution seismic reflection profiles of the insular shelf off of western Irabu Island also show the existence of mound structures located at present in water depth of ca. 126 m that were interpreted as reefal facies (Obata and Tsuji 1992). One of these was drilled, and its carbonate sediments were analyzed in detail; some coral fragments were dated by the 230Th/234U and 14C methods (Sasaki et al. 2006). This lithological examination and identification of coral fauna confirmed that the upper part of the mound structure is a small-scale coral reef, which was likely deposited until ca. 22 ka during the last glacial period.

4.4 Natural Environment of the Late Pleistocene Human Subsistence

Most intriguing from the perspective of landscape history is the possibility of coral reef formation during even the last glacial maximum (LGM) because the Ryukyu Island Arc had already been colonized by groups of Pleistocene humans in this period. Fossilized human remains have been excavated from calcareous fissure and cave sediments of limestone areas on Okinawa and Kume Islands in the central Ryukyu and also on Miyako and Ishigaki Islands in the southern Ryukyu. Many of the discovered Pleistocene human remains from Japan are concentrated within this arc (Matsu'ura 1996). So far, the earliest radiometric age for these Pleistocene human remains is ca. 32 ka from Yamashita-cho Cave 1 of Okinawa Island, where a juvenile femur and tibia were excavated in association with abundant extinct deer (*Cervus astylodon*). The layer including these remains was covered with a deposit containing dated charcoal flecks, probably from an earth oven; thus, the layer coincides with or is older than the conventional age of 32,100 ± 1000 BP (Takamiya et al. 1975).

Human skeletal remains of more than five individuals were excavated at a fissure site of the Minatogawa limestone quarry on Okinawa Island (Suzuki and Hanihara 1982). The lower deposits of this fissure also yielded fossils of extinct deer and less abundant wild boar (*Sus leucomystax taivanus*), whereas wild boar fossils were found without deer in the upper deposits. To confirm the vertical, or chronological, provenance of the human remains, a comparative analysis of human and terrestrial vertebrates was applied to their fluorine contents taken from post-burial surrounding sediments (Matsu'ura 1996). The result indicated that the human remains are close in relative age to the vertebrate bones from the lower deposits. Two conventional ages of 18,250 ± 650 and 16,600 ± 300 BP, respectively, were also obtained from charcoal flecks from the lower deposits (Suzuki and Hanihara 1982).

Most of the ages are less reliable ones from charcoal flecks collected from associated deposits of the fissure and cave sites, and the possibility remains that these could be contaminated by secondary deposits (Matsu'ura 1996). However, some exceptional human remains excavated with terrestrial vertebrates (including wild boar) at Shiraho-Saonetabaru Cave on Ishigaki Island, the southern Ryukyu, have recently provided more reliable conventional ages from well-preserved collagen samples extracted directly from the bones: 20,416 ± 113 and 18,752 ± 100 BP (Nakagawa et al. 2010). These ages are just within the LGM, when the lowstand sea level should have been at least 120 m lower than the present sea level. The islands at that time were larger and higher than the present ones, and some of them, such as Ishigaki and Iriomote, were likely connected into a "super island" (Ota 1998:189). A paleobotanical analysis of a boring core sample from Izena Island in the central Ryukyu indicated that the pollen assemblages around ca. 22 ka are dominated by coniferous trees, such as *Pinus* and *Podocarpus* with substantially less *Abies* and *Tsuga*, and deciduous broad-leaved trees of *Alnus* and *Quercus* subgen. *Lepidobalanus*. Thus, it has been assumed that the last glacial climatic condition in the subtropical Ryukyu Islands was not much cooler but was more arid than the climate of the present day (Kuroda and Ozawa 1996). This assemblage also included considerable pollen of evergreen broad-leaved trees, such as *Quercus* subgen. *Cyclobalanopsis*, *Castanopsis-Pasania*, and *Myrica*. This nuciferous vegetation would have supported terrestrial vertebrates such as deer and wild boar.

The late Pleistocene human groups in the Ryukyu Island Arc were likely dependent on nuts and the terrestrial vertebrates living within the evergreen broad-leaved forests. This assumption appears to be supported by the fossilized bones of deer and/or wild boar excavated along with human remains from fissure and cave sites located at the far interior of super islands, which would be within the habitats of these vertebrates. Nevertheless, recent geological investigations of some island shelves suggest that a shallow-water environment with scattered small coral reefs was also available to the Pleistocene human groups, even in the period of the LGM. This new possibility deserves to be examined archaeologically and anthropologically.

4.5 Coral Studies Linking the Late Pleistocene to the Holocene

During the post-LGM, the relative sea-level change has been dominated by a transition of the earth system from a glacial to an interglacial stage, and the large-scale ice-to-ocean mass redistribution has produced a dramatic increase in ocean water volume as well as a large and ongoing isostatic response of the solid Earth (Fleming et al. 1998; Milne et al. 2005). According to the sea-level history obtained by synthesizing three deglaciation curves based on long drill cores from Tahiti, Barbados, and New Guinea in the far-field region, where the glacial isostatic adjustment should be smaller in magnitude and the eustatic signal therefore dominant, there are two brief periods of accelerated ice melting, at ~13.8 and 11.3 ka, respectively, and these so-called meltwater pulse events are thought to have induced rapid sea-level rises that drowned the late Pleistocene coral reefs, which could not keep up with the rising water level (Camoin et al. 2005).

The present coral reefs surrounding the islands of the southern and central Ryukyu Arc started to grow some tens of meters below the present sea level since ca. 8.5–7.9 ka, but this seems to have occurred more than 1500 years later in the northern islands (Kan and Kawana 2006). The global melting rate of ice sheets has been significantly decreased from around 7 cal. kyear BP to the late Holocene (Milne et al. 2005); thus, the smoother sea-level rise should have been a suitable environment for coral reef growth. Additionally, the main axis of the Kuroshio, a warm current, is considered to have reentered the Okinawa Trough around 7.3 cal. kyear BP, when the sea surface temperature and depth of the thermocline abruptly increased to higher than modern values in both southern and northern troughs (Jian et al. 2000). The warm water mass transported from the tropics would also have increased the vertical growth rate of coral reefs in the Ryukyu Island Arc (Kan 2010:24).

Studies have illustrated the drastic changes in the insular environment from the late Pleistocene to the mid-Holocene. In particular, portable submersible drilling devices, newly developed in the 1970s (e.g., Macintyre 1975), enabled marine geologists to directly observe the interior coral reefs; description of their developmental processes was also facilitated by the prevalence of the radiocarbon dating method (Kan 2002; Hongo 2010). Consequently, it has been demonstrated that the Holocene coral reefs caught up with sea level during the mid-Holocene, which should be crucial to archaeological investigation of the Neolithic hunting-and-gathering subsistence economy and habitation in the insular environment of the Ryukyu Arc. For instance, as the coral reefs developed and their wave-dissipating effect increased, the Holocene high-energy window directly eroding island coasts would be closed, and the shallow lagoon environment could then form (Kan 2002). The prehistoric subsistence strategy should have been influenced by such a change in the coastal environment, resulting in a much greater dependence on marine resources available from the shallow water of the coral lagoon (e.g., Kinoshita 2012:17–18; Toizumi 2010:16).

However, considerable differences exist among the "catch-up" chronologies of the coral reefs, which range from 6.5 to 3.5 ka (Kan 2010). Even at the same island, the windward reef generally developed earlier than the leeward reef (Hongo and Kayanne 2008). This chronological diversity can be attributed to local effects of tectonic movement, wind direction, wave strength, and sea surface temperature. The spatial distribution of archaeological sites should thus be reexamined in detail with the local chronologies of Holocene coral reef development. Although Neolithic subsistence in the mid-Holocene in the Ryukyu Island Arc is generally viewed to have been largely dependent on the coral reef ecosystem, a recent case study of Ishigaki Island, in the southern Ryukyu, showed that the locations of habitation sites are not exclusively restricted within the spatial range of developed coral reef landforms in either the Shimotabaru (ca. 4850–3640 cal. BP) or Mudoki (ca. 1880–930 cal. BP) periods (Kobayashi et al. 2013)[1].

The recent coral studies can provide a diachronic perspective on environmental history, linking the late Pleistocene to the Holocene. These have revealed drastic changes in the insular environment, which are crucial to and must be considered in archaeological investigations and interpretations, as mentioned previously. Similarly, the consequent relative sea-level fall in the late Holocene should have changed the littoral environment. The sea-level fall and seaward accretion of the reef should have resulted in reduced wave force at the shore. Such wave attenuation would bring about changes in the sedimentary environment and also promote sand spits or bars on emergent reefs. In the case of Tonaki Island in the central Ryukyu, a prehistoric habitation site dated at ca. 3500 BP and containing earthenware and shell middens (Toma 1981) is located on a tombolo, a sand bar connecting two islands (Kan et al. 1997).

The above archaeological site indicates that the geomorphic accretion of the littoral environment enlarged the spatial range of human habitation. However, human activities, even in prehistoric time, have another aspect: they act as an agent affecting the environment. Thus, we need to consider the anthropogenic changes in the landscape as well as the adaptive aspect of prehistoric subsistence. This requires the articulation of two directions of investigation in the diachronic perspective, the geomorphology of shallow water, where corals have been distributed, and the geoarchaeology of land, where islanders have lived. Our research of the Nagura region, Ishigaki Island, is still in progress, but a preliminary outline is presented below.

4.6 The Possibility of Anthropogenic Landscape Changes in the Late Holocene

Ishigaki Island, one of the southern Ryukyu Islands, lies about 440 km southwest of Okinawa Island and 240 km east of Taiwan. It has an area of approximately 223 km2, a long axis of 32 km extending from northeast to southwest, and a maximum width of 19 km at its southwestern part, where the Nagura region is located, facing the shallow Nagura Bay (Fig. 4.1b). This region is a large alluvial fan of the Nagura River, which is located between two chains of mountain peaks: the southern side of Banna-dake and the northern side of Omoto-dake, which is the highest peak at 526 m. The lower terrace of the nearly level or rolling alluvial fan consists of light-red to yellowish-red loamy matrix with lenses and beds of clay, sand, cobbles, and boulders (Foster 1965: Geologic Map of Ishigaki-Shima, Ryukyu-Retto). Ranging from 5 to 15 m above sea level, it slopes toward the bay; it is bounded by the alluvial lowland and is dissected by the main stream and several small valleys. In front of the alluvial lowland is an extensive mangrove forest of tideland, which is separated from the tidal flat of Nagura Bay by some spits covered with coastal vegetation.

In both mangrove tideland and tidal flat, there are many scattered emergent microatolls of *Porites* and *Favia* (Fig. 4.2). During the low tide of 2012 spring tides, we measured the horizontal and vertical distributions of microatolls observable in the channels and lagoons of the tideland, including those on a transect line between the sand spit and the edge of the tidal flat, using the RTK-GPS method. We also collected some small samples for dating analyses. Although the microatolls should have been eroded to some degree, it was discernible that the rear microatolls in the tideland were higher, and the ones nearer to the edge of the tidal flat were lower; the heights of their flat surfaces ranged from ca. +1.5 to +0.5 m above the present mean low water level (LWL). Some dated samples, collected from

[1] Conventionally, the prehistory of Ishigaki Island in the Holocene has been divided into two cultural phases in accordance with archaeological assemblages, along with a cultural blank between these phases: the Shimotabaru (the earlier phase) and the Mudoki (the later phase).

Fig. 4.2 We measured emergent *Porites* and *Favia* microatolls (*yellow dots*) in the mangrove tideland and along a transect in the tidal flat of the Nagura region by the RTK-GPS method. Several colonies of dead *Porites* (*yellow crosses*) were found lying at a depth of about 2 m

surface edges of the microatolls, showed a time difference of ca. 3000 years, i.e., between 5000 and 2000 cal. BP, with rear ones being earlier than the forward ones (Fig. 4.3). Decreases in the heights and ages of the microatolls would be associated with the relative fall of sea level during the late Holocene, but we should also consider the uplift caused by tilting of southern Ishigaki in further interpretations (Kawana 1989). These emergent microatolls seem to have trapped soil transported from inland. Consequently, a shallow sea of microatolls drastically changed its landscape into a tideland of mangrove and a tidal flat during the late Holocene.

We also observed that several colonies of dead *Porites*, which had diameters of several meters, were scattered at the sandy bottom of Nagura Bay at a depth of approximately 2 m. Two dated samples from these suggested that their growth was restrained at about 800–1200 cal. BP. Their decline could not be attributed to sea-level change but may be related to other environmental factors. Although it is necessary to examine such factors by analyses of trace elements in coral skeletal cores[2], geoarchaeological information from the alluvial lowland will also be helpful for reconstructing the landscape history, including aspects of both land and shallow sea.

We collected boring cores on the Nagura coastal lowland and within the bottom of a small valley bisecting the lower terrace. The latter core, particularly, provided a stratigraphic sequence from shallow sea, through mangrove tideland to alluvial lowland, which subsequently has been altered into irrigated pond fields for rice cultivation. Floodplain deposits of sandy silt between sediments of tideland and pond fields accumulated to a thickness of more than 50 cm in the valley bottom around 2000–1000 cal. BP and contain a large number of *Pundanus* and *Barringtonia* pollen, which increase rapidly upward. Because its silica (SiO_2) content is increased compared with lower sediments, the deposit can be viewed as having been derived from inland terrace. This phase corresponds to a period of prehistoric cultural, and some archaeological sites have been discovered within the Nagura region: the Kanda habitation site at the skirt of the lower terrace and shell midden spots of Nagura Site on an early

[2] A coral skeletal core was bored in a colony of dead *Porites* and dated at ca. 1200 cal. BP, which has a time window of 10 years. The latest analysis of its Ba/Ca (a proxy of sedimentation and nutrients) shows no cyclical peaks suggesting seasonal activities, such as agricultural activity (Sowa et al. 2014).

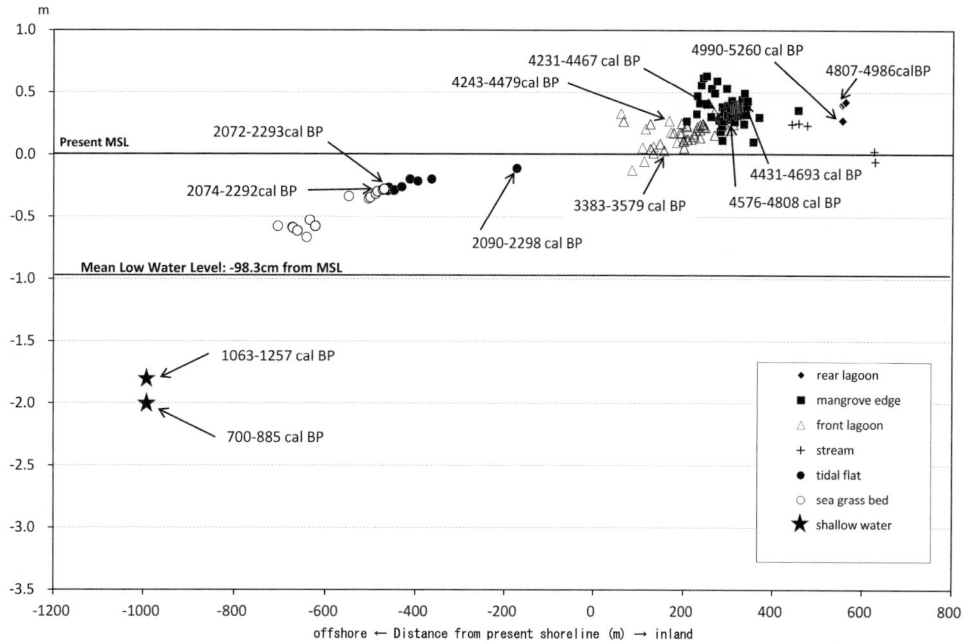

Fig. 4.3 Vertical distribution of emergent *Porites* and *Favia* microatolls and calibrated ages of collected samples. Decreases in the heights and ages of the microatolls would be associated with the relative fall of sea level during the late Holocene

sand spit. In the boring cores, microscopic charcoal flecks are more abundant within the sandy silt than in the lower sediments, which suggests the possibility that anthropogenic ignition of broad-leaved forest occurred synchronously with deposition of the floodplain deposit. Although detailed analyses are still in progress, it seems possible that colluvial soil washed from the inland not only contributed to the stabilization of an alluvial environment suitable for irrigation but also likely degraded the coastal shallow-sea environment where *Porites* and *Favia* colonies grow, as suggested by recent damage to coral growth by outflows of inland laterite soil.

4.7 Concluding Remarks and Introduction to the Next Two Chapters

Coral studies of the Ryukyu Island Arc have resulted in the accumulation of a vast knowledge quite relevant to a future reconstruction of the long-term landscape history, which can be articulated with anthropological and archaeological evidence on late Pleistocene and Holocene human activities in the insular environment. The preliminary articulation of several colonies of dead *Porites* with data from boring cores from a small valley suggests two aspects of anthropogenic work: environmental degradation and landscape enhancement. The first refers to the fact that inland activities degraded the shallow-water environment of coral colonies, even in prehistoric time, and the second focuses on the notion that these activities also promoted the development and stabilization of the alluvial environment in small valleys that were consequently altered into irrigated pond fields for rice cultivation. The detailed analysis of landscape history will reveal the difficulty of distinguishing between degradation and enhancement; thus, the essential aim should be to illustrate interactions between two agents, natural and human, through a cumulative and contingent history of the landscape (Kirch 2005:432) from the distant past to the present and toward the future.

This perspective of discerning landscapes as historical products is certainly not unique, as it was shared among the natural sciences and humanities as early as A.D. 1955. In that year, the Wenner-Gren Foundation for Anthropological Research held an international symposium on "Man's Role in Changing the Face of the Earth" at Princeton University, which was a milestone event at which researchers from several academic domains such as geography, ecology, zoology, botany, history, anthropology, and planning came together to discuss human-induced environmental changes (Thomas 1956). Carl Sauer, a noted geographer, led the organization of this symposium to consider a tripartite division: retrospect, process, and prospect. This approach considers that retrospect and prospect are different ends of the same sequence and that today is therefore but a point on a line (Williams 1987:225). The approach of landscape history should also be open to discussions of future landscapes produced by on-going interactions between the nature and humans.

A recent work in planning has also appealed to the importance of orienting the present and future landscape through a diachronic perspective (Marcucci 2000). It allows us to summarize and complement the effectiveness of this perspective as follows: (1) understanding the long-term changes in landscape moderates the narrow essentialism of admiring specific cultural traditions, pristine landscapes, or harmonized ecosystems, which appear to stand on the edge of a precipice; (2) concurrence in the dynamic interactions between humans and nature is useful for overcoming the dichotomy between fundamental environmentalism and extreme development; (3) the diachronic historical perspective aids our realization that present planning becomes part of the historical process of a landscape once it is implemented; and (4) initiating historical inquiry at the very beginning provides a means for engaging the local inhabitants in landscape planning and also allows them to be experts on the history of their landscape.

The fourth point, particularly, would be a unique conception for planning, and it suggests that archaeologists can also contribute by weaving academic outcomes and the narratives of personal histories into a local landscape history, which would promote dialogue among inhabitants, including those who are not necessarily concerned with local environmental issues, in various situations. For such a contribution, we need to familiarize ourselves with historical–anthropological studies of the recent past, in which the present inhabitants have lived, and consider ways of making use of landscape histories in their present and future lives. The next two papers will include many suggestions on these topics. One of these papers is on the disappearance of historical ecological knowledge associated with destructive coral mining for traditional plaster in Ishigaki (Fukayama), and the other is on the practical articulation of local or personal knowledge for exploiting living resources and academic outcomes of landscape history in a modern enterprise such as ecotourism (Shimoda). The first suggests that traditions do not necessarily fit in with modern environmental conservationism but that they should rather be viewed as ecological and local knowledge about living resources. The second proposes a way of thinking, not about protecting corals and coral reefs, but about inventing value in them, both economic and cultural.

References

Camoin GF, Yasufumi I, McInroy D (2005) The last deglacial sea level'rise in the South Pacific: offshore drilling in Tahiti (French Polynesia). IODP Sci Prosp 310:1–51. doi:10.2204/iodp.sp.310.2005

Fleming K, Johnston P, Zwartz D, Yokoyama Y, Lambeck K, Chappell J (1998) Refining the eustatic sea-level curve since the last glacial maximum using far- and intermediate-field sites. Earth Planet Sci Lett 163:327–342

Foster HL (1965) Geology of Ishigaki-shima, Ryukyu-retto, Geological Survey professional paper 388-A. US Government Printing Office, Washington, DC

Hikida T, Ota H (1997) Biogeography of reptiles in the subtropical East Asian islands. In: Proceeding of the symposium on the phylogeny, biogeography and conservation of Fauna and Flora of East Asian Region, National Science Council, R.O.C, pp 11–28

Hongo C (2010) High-resolution Holocene sea-level change based on coral reefs and hermatypic corals. J Geogr (Chigaku Zasshi) 119 (1):1–16 (In Japanese)

Hongo C, Kayanne H (2008) Holocene coral reef development under windward and leeward locations at Ishigaki island, Ryukyu islands, Japan. Sediment Geol 214:62–73

Ikeda S, Kasuya M, Ikeya M (1991) ESR ages of middle Pleistocene corals from the Ryukyu islands. Quat Sci 36:61–71

Jian Z, Wang P, Saito Y, Wang J, Pflaumann U, Oba T, Cheng X (2000) Holocene variability of the Kuroshio Current in the Okinawa Trough, northwestern Pacific Ocean. Earth Planet Sci Lett 184:305–319

Thomas WL Jr (ed) (1956) Man's role in changing the face of the earth: an international symposium under the co-chairmanship of Carl O. Sauer, Marston Bates, Lewis Mumford. Univ. of Chicago Press, Chicago

Kan H (2002) Development of coral reefs and the environmental change in coastal zone – sea level rise and revisiting of Holocene high energy window. Bull Soc Sea Water Sci Jpn (Nippon Kaisuigakkaishi) 56(2):123–127 (In Japanese)

Kan H (2010) Holocene history coral reefs in the Ryukyu islands. Archaeol J (Koukogaku J) 597:24–26 (In Japanese)

Kan H, Kawana T (2006) Catch-up of a high-latitude barrier reef by back-reef growth during post-glacial sea-level rise, Southern Ryukyus, Japan. In: Proceedings of the 10th international coral reef symposium, Japanese Coral Reef Society, Tokyo, 494–503

Kan H, Hori N, Kawana T, Kaigara T, Ichikawa K (1997) The evolution of a holocene fringing reef and island: reefal environmental sequence and sea level change in Tonaki island, the central Ryukyus. Atoll Res Bull 443:1–20

Kawana T (1989) The Quaternary changes of earth's crust in the Ryukyu islands. Chikyu Mon (Gekkan Chikyu) 11(10):618–630 (In Japanese)

Kinoshita N (2012) Formation of culture and movement of people in prehistoric Ryukyu-with focus on the mutual human- geographical relation between islands. Bungakubu-ronso (Univ of Kumamoto) 103:13–27 (In Japanese)

Kirch PV (2005) Archaeology and global change: the Holocene record. Annu Rev Environ Resour 30:409–440

Kirch PV, Hunt TL (eds) (1997) Historical ecology in the Pacific islands. Yale University Press, New Haven and London

Kizaki K (1980) Ryukyu-ko no Chishitsu-shi. Okinawa Times Press, Naha (In Japanese)

Koba M (1992) Influx of the Kuroshio current into the Okinawa trough and inauguration of quaternary coral reef building in the Ryukyu island arc Japan. Quat Res (Dai Yonki Kenkyu) 31 (5):359–373

Kobayashi R, Yamaguchi T, Yamano H (2013) Remote sensing evaluation of spatio-temporal variation in coral-reef formation in Ishigaki, Yaeyama islands, toward a study of prehistoric insular resource utilization. Q Archaeol Stud 60(2):55–72 (In Japanese)

Kuroda T, Ozawa T (1996) Studies of land bridges and the migration of men and other animals along them – paleoclimatic and vegetational changes during the Pleistocene and Holocene in the Ryukyu islands inferred from pollen assemblages. J Geogr (Chigaku Zasshi) 105 (3):328–342 (In Japanese)

Machida H et al (eds) (2001) Nippon no Chikei 7: Kyushu and Nansei-shoto. University of Tokyo Press, Tokyo (In Japanese)

Macintyre IG (1975) A diver-operated hydraulic drill for coring submerged substrates. Atoll Res Bull 185:21–26

Marcucci DJ (2000) Landscape history as a planning tool. Landsc Urban Plan 49:67–81

Matsuura S (1996) A chronological review of Pleistocene human remains from the Japanese Archipelago. In Omoto K (ed) Interdisciplinary perspectives on the origins of the Japanese, International Research Center for Japanese Studies, Kyoto, 181–197

Matsuda H, Arai K, Machiyama H, Iryu Y, Tsuji Y (2011) Submerged reefal deposits near a present-day northern limit of coral reef formation in the northern Ryukyu island arc, northwestern Pacific Ocean. Island Arc 20:411–425

Matsuda H et al (2012) Paleoceanography reconstructed from shallow-water carbonates during glacial periods in the northern part of the Ryukyus. Chikyu Mon (Gekkan Chikyu) 34(6):363–372 (In Japanese)

Matsui M, Ito H, Shimada T, Ota H, Saidapur SK, Khonsue W, Tanaka-Ueno T, Wu G (2005) Taxonomic relationships within the Pan-Oriental Narrow-mouth Toad *Microhyla ornata* as revealed by mtDNA analysis (Amphibia, Anura, Microhylidae). Zool Sci 22:489–495

Milne GA, Long AJ, Bassett SE (2005) Modelling Holocene relative sea-level observations from the Caribbean and South America. Quat Sci Rev 24:1183–1202

Nakagawa H (1983) Outline of Cenozoic history of the Ryukyu islands. Geol Soc Jpn (Chishitsugaku-ronsyu) 22:67–79 (In Japanese)

Nakagawa R, Doi N, Nishioka Y, Nunami S, Yamaguchi H, Fujita M, Yamazaki S, Yamamoto M, Katagiri C, Mukai H, Matsuzaki H, Gakuhari T, Takigami M, Yoneda M (2010) Pleistocene human remains from Shiraho-Saonetabaru Cave on Ishigaki island, Okinawa, Japan, and their radiocarbon dating. Anthropol Sci J Anthropol Soc Nippon 118(3):173–183

Obata M, Tsui Y (1992) Quaternary geohistory inferred by seismic stratigraphy of a carbonate province in an active margin, off Miyako island, south Ryukyus, Japan. Carbonates Evaporites 7:150–165

Osozawa S, Shinjo R, Armid A, Watanabe Y, Horiguchi T, Wakabayashi J (2012) Palaeogeographic reconstruction of the 1.55 Ma synchronous isolation of the Ryukyu islands, Japan, and Taiwan and inflow of the Kuroshio warm current. Int Geol Rev 54 (12):1369–1388

Ota H (1998) Geographic pattern of endemism and speciation in amphibians and reptiles of the Ryukyu Archipelago, Japan, with special reference to their paleogeographical implications. Res Popul Ecol 40(2):189–204

Ota H (2003) Toward a synthesis of paleontological and neontological information on the terrestrial vertebrates of the Ryukyu Archipelago, I: systematic and biogeographic review. J Fossil Res (Kaseki Kenkyu-kai Kaishi) 36(2):43–59

Sagawa N, Nakamori T, Iryu Y (2001) Pleistocene reef development in the southwest Ryukyu islands, Japan. Palaeogeogr Palaeoclimatol Palaeoecol 175:303–323

Sasaki K, Omura A, Miwa T, Tsuji Y, Matsuda H, Nakamori T, Iryu Y, Yamada T, Sato Y, Nakagawa H (2006) ^{230}Th/^{234}U and ^{14}C dating of a lowstand coral reef beneath the insular shelf off Irabu island, Ryukyus, southwestern Japan. Island Arc 15:455–467

Sowa K, Watanabe T, Kan H, Yamano H (2014) Influence of land development on Holocene *Porites* coral calcification at Nagura bay, Ishigaki island, Japan. PLoS One 9(2):e88790. doi:10.1371/journal.pone

Suzuki H, Hanihara K (eds) (1982) The Minatogawa man: the upper Pleistocene man from the island of Okinawa, Bulletin, no. 19. University Museum, University of Tokyo, Tokyo, Japan

Takamiya H, Kin M, Suzuki M (1975) Excavation report of the Yamashita-cho Cave Site, Naha-shi, Okinawa. J Anthropol Soc Jpn (Jinruigaku Zasshi) 83(2):125–130 (In Japanese)

Toizumi T (2010) Vertebrate resource use in prehistoric Amami and Okinawa islands. Archaeol J (Koukogaku J) 597:15–17 (In Japanese)

Toma S (1981) Tonaki-jima no Iseki. Reports of Okinawa Prefectural Museum II (Okinawa Kenritsu Hakubutsukan Sogo Chosa Hokokusho II), Tonaki Island, pp 41–49 (In Japanese)

Williams MW (1987) Sauer and "Man's role in changing the face of the earth". Geogr Rev 77(2):218–231

Yamaguchi T, Kayanne H, Yamano H (2009) Archaeological investigation of the landscape history of an Oceanic Atoll: Majuro, Marshall Islands. Pac Sci 63(4):537–565

Producing Plaster: Traditional Uses and Knowledge of Coral on Ishigaki Island, Okinawa

5

Naoko Fukayama

Abstract

A lagoon reclamation project on Ishigaki Island for the purpose of building a new airport caused much controversy at the end of the 1970s because it would destroy the well-developed coral reefs around the island. Since then, the view that corals were biologically and ecologically valuable and to be conserved has pervaded the island. During this period, however, it has been almost forgotten that corals used to be used as building materials, in particular, lime plaster—a highly processed product derived from corals. From fieldwork, particularly interviews with retired plasterers, I reconstruct the process of traditional plaster production. I also discuss why and how plastering flourished and declined between the 1940s and the 1960s and describe how social attributes of the plasterers on the island differed from those on Okinawa Island—the main island of the Ryūkyū Arc—from which the art of plastering was imported. To consider conservation and recovery of coral at the local level today, it is important to respect such a history of coral uses and to recognize the possibility of various concerns about coral.

Keywords

Plaster • Plasterer • Traditional uses and knowledge

5.1 Introduction

Ishigaki Island (石垣島, Ishigaki-jima) is the second-largest island in the Yaeyama Islands of the southern Ryūkyū Arc. It is located more than 400 km from Okinawa Island—the main island of the Ryūkyū Arc—but less than 300 km from Taiwan. The island is administratively within the city of Ishigaki in Okinawa Prefecture and is the center of business and transport in the Yaeyama Islands. The population of the city is about 47,000 at present.

The islands in the Ryūkyū Arc are within a subtropical climate zone and have unique terrestrial and aquatic environments. Well-developed coral reefs can be observed, especially around the Yaeyama Islands, Miyako Islands, Okinawa Island, and Amami Islands. They have attracted a number of domestic and international tourists. However, these coral reefs have been degraded over decades as a result of anthropogenic effects, such as land development, especially after World War II. Moreover in the 1970s and 1980s, they were severely damaged by crown-of-thorns starfish. Then, in 1998, coral bleaching also occurred widely, due to the increase in seawater temperatures.

Since these two events, conservation and recovery of corals have been recognized as an urgent issue, and many scientists and nature conservationists have visited the islands to survey the present state of the corals and discuss countermeasures. Indeed, it seems that something of a "coral boom" has occurred and that a certain view of the corals, that corals are biologically and ecologically valuable and to be conserved, has come to be emphasized on the islands and more widely.

N. Fukayama (✉)
Department of Social Anthropology, Tokyo Metropolitan University, 1-1, Minami-osawa, Hachioji-shi, Tokyo 192-0397, Japan
e-mail: fukayama@tmu.ac.jp

H. Kayanne (ed.), *Coral Reef Science*, Coral Reefs of the World 5, DOI 10.1007/978-4-431-54364-0_5

Against that, at the end of the 1970s, Okinawa Prefecture made a plan for the reclamation of a lagoon in Shiraho, located in the southeast of Ishigaki Island, to build a new airport. This project caused much controversy among scientists and nature conservationists from outside the area, as well as among local inhabitants. Shiraho is famous for its rich corals, particularly for large colonies of blue corals, and corals became the most powerful motivation for opposing the reclamation. The view that corals were biologically and ecologically valuable and to be conserved pervaded the discussion. The cultural value of corals was also invoked to support this view, and it was emphasized that Shiraho's lagoon had traditionally been a local fishing ground, on a daily basis (e.g., Noike 1990; Kobashigawa and Mezaki 1989).

The long-standing discussion did bear fruit; eventually, in 1989, the reclamation plan was abandoned. In fact, the new airport issue seemed to have accelerated the "coral boom" on Ishigaki Island.

I started my fieldwork in 2009 with an interest in the position of corals and coral reefs in present-day local communities. In my fieldwork, I soon noticed that a "positive" view of corals was not necessarily shared deeply among the local inhabitants, particularly among those who had experienced the era before the "coral boom." Indeed, some of them told me that they hardly ever went to the beaches or the sea and knew almost nothing about the corals. One man in his 80s pointed out that it was only recently that people started to say anything about the corals. Another inhabitant, in his 70s, who had been engaged in fishing, commented that there used to be nobody who valued the corals as good or pretty objects, but that the situation had changed since the new airport issue had emerged. Thus, I came to doubt that the "positive" view of corals and coral reefs was local or traditional.

In fact, according to a classic ethnography, written by a local ethnographer Fumi Miyagi, fishing on the island had long been "primitive" and "very inactive," and functioned only on a minor subsistence basis. She did mention, however, the direct use of corals as building materials and for kitchen wares (Miyagi 1982 (1972)). Furthermore, a reliable dictionary of the Yaeyama dialect, written by Tousou Miyara, a local linguist, included words related to coral. Table 5.1 provide a list of them and show the possibility that corals were, at least in 1930, recognized basically as "stones" among the original inhabitants using the local dialect (Table 5.1) (Miyara 1930).

Thus, there would seem to be a gap between the present view and the old, local view regarding the corals. Bearing this in mind, in this chapter, I provide a short review of lime

Table 5.1 Yaeyama dialect words related to corals from "Yaeyama vocabulary" (Miyara 1930)

Words	Specific plane	Phonetic symbols	Part of speech	Literal meaning
イナガ・イシ	Shiraho	inaga-is'i	n.	Sea stone
イン・マチィ	Ishigaki, Shiraho	im-matsï	n.	Sea pine
ウール	Ishigaki	u:ru	n.	Sea stone
ウル・イシ	Ishigaki, Shiraho	uru-is'i	n.	Sea stone
カサ・イシ	Ishigaki	kasa-is'i	n.	Shade stone
カツォーラ・イシ	Shiraho	katso:ra-is'i	n.	Shade stone
チィブリィ・イシ	Ishigaki	tsïburï-is'i	n.	Chrysanthemum stone
ボージィ・イシ	Ishigaki	bo:dzï-is'i	n.	Shaven head stone
ムン・ッシィ・イシ	Ishigaki	muɲ-ssï-is'i	n.	Wheat thrashing stone
ウール・ヌ・パイ	Ishigaki	u:ru-nu-pai	n.	Sea stone ash
ウールヌパイ・イシ	Ishigaki	u:ru-nu-pai-is'i	n.	Stone for sea stone ash

*Here, "Ishigaki" means former Ishigaki-cho, the western half of the Ishigaki Island. Miyara wrote that the standard Yaeyama language was spoken in Ishigaki-cho (Miyara 1930: 3)

plaster—a highly processed product derived from corals—and plastering, from fieldwork I conducted intermittently from 2009 to 2012 on Ishigaki Island.

5.2 History of Plaster in Okinawa

If you walk on the island, especially around the areas that have been populated for generations, such as the urban district known as Shika, you can easily find corals used as building materials at private houses (Fig. 5.1). Typically, there are foundation stones of brain corals, with their round shape, curbstones of hump corals that are easy to cut, and spread gravel of branch corals. Hump corals are also used as wall blocks. Sacred and religious places, called "on" in the Yaeyama dialect—"utaki" in the standard Okinawan dialect—are often full of coral materials, and, sometimes, corals themselves are treated as objects of worship (Fig. 5.1). Apart from building materials, corals have also been used for kitchen wares, such as grinders, because of their jagged surfaces, as Miyagi noted, and also as weights for looms or boats, just like stones.

Today, it is almost impossible for the local inhabitants to extract corals from the lagoons, as I discuss below, but in the past, inhabitants used to gather living corals when they needed to process them because "raw" ones were softer. They also gathered dead corals, mainly on the beaches, or reused old materials if they required little processing.

Lime plaster, or simply plaster, on which I focus here, is the most processed coral product in Okinawa. It is well

Fig. 5.1 Corals used as building materials at a private house and at a sacred and religious place on Ishigaki Island, August 2009 (Photo by author)

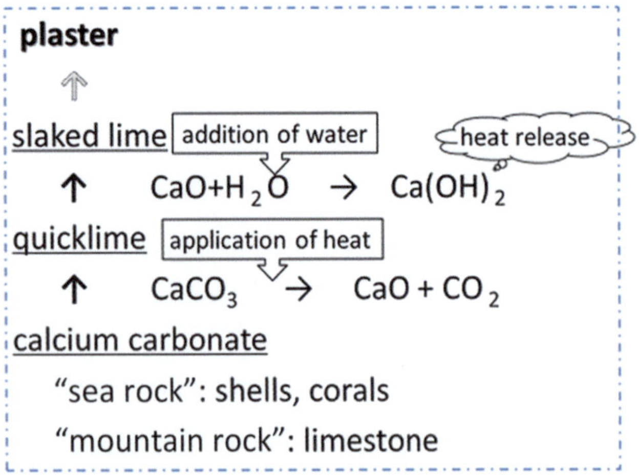

Fig. 5.2 Main ingredients of plaster

known that plaster as a building material has been used in various places all around the world since ancient times, mainly for making flat and smooth surfaces for walls and floors. Lime plaster is one type; the paste, basically composed of slaked lime and water, hardens as it dries.

The raw materials for slaked lime in Japan can be classified roughly into "mountain rocks", that is, limestone, and "sea rocks", including shells and corals. While "mountain rocks" are much more common, the primary component of both is the same, calcium carbonate. During processing, firing (calcination) converts "rocks" into quicklime (calcium oxide, CaO), and subsequent addition of water converts the quicklime into slaked lime or hydrated lime (calcium hydroxide, $Ca(OH)_2$), with the release of heat (Fig. 5.2).

Corals had been used as a raw material for plaster not only in Okinawa but also in several areas in the Shikoku and Kyūsyū regions of southeastern Japan. Usually chopped rice straw was also added to the paste in Okinawa, instead of seaweed and hemp, as in many other regions in Japan. The final product is called "muchi" in Okinawa, including Ishigaki Island, which originally meant pounded rice cake, because the plaster has a similar texture. Powdery quicklime is also called, at least in Shika of Ishigaki Island, "ūru-nu-pai," which literally means ash of coral or "sea rocks" (Table 5.1) (Miyara 1930).

Plaster was used primarily for roofing in Okinawa, not for smoothing walls or floors. Traditional roofs are often covered with unglazed tiles, called "kawara," and plaster is applied to the joints to reinforce the whole roof, in particular to resist heavy rainstorms in the typhoon season. It was important to apply plaster thoroughly. Plaster was also used to shape tombs and to join boat planks and wall blocks. It is also well known that quicklime, before being made into slaked lime, was an indispensable additive in producing brown cane sugar.

There are few remaining historical documents on plaster works in Okinawa. According to the description of "Kyūyō," a Chinese man who drifted ashore was the one who propagated the manufacturing method for firing shells in kilns in 1731 (Kyūyō Kenkyūkai 1974: 298; Kuniyoshi 2004: 41). However, it is also known that plaster was applied to the stone walls of Shuri Castle, the palace of the Ryūkyū Kingdom, built in the fifteenth century AD. Thus, the use of kilns presumably started earlier than that (Kuniyoshi 2004: 41). After they started manufacturing plaster, the kingdom came to control that work under the magistrate's office of roof tiles. During this time, roofing with tiles was only allowed at Shuri Castle, temples, houses of certain "shizoku" (social class of rulers), and village offices, while thatch roofs were much more common. Thus, the use of roof tiles was centered on the cities of Shuri and adjacent Naha. It is assumed that the demand for plaster was high only in these cities; that is, construction with roof tiles and plaster used to be an urban phenomenon.

In 1879, the Meiji Japanese government abolished the Ryūkyū Kingdom when the islands were incorporated as Okinawa Prefecture. Any restriction on the use of roof tiles was practically lifted then, and demand for plaster among the people generally increased (Ishii 2010: 182).

During World War II, many buildings were destroyed, especially in the Battle of Okinawa. After the war, constructing buildings and roofing with tiles were encouraged in the process of recovery, led by the government of the Ryūkyū Islands—that is, self-government by native Okinawans during the American occupation of Okinawa. The "construction rush" had started and again the production of plaster became a viable business. However, shortly thereafter new materials, such as industrial cement, were introduced, and new building styles—without tiles—were also invented. Thus, the demand for plaster decreased rapidly. In 1972, when Okinawa reverted from the USA to Japan, "coral fishing," that is, fracturing reefs, and collecting sand and rocks were all banned under the Sea Area Fisheries Adjustment Regulations (Okinawa Prefecture website).

On Ishigaki Island, it is recorded that roof tile craftsmen were brought from Okinawa Island and they started manufacturing tiles in a kiln in 1695 (Editorial Section of the General Affair Department, Ishigaki City 1999: 41). There is hardly any documentation regarding plaster then, but it seems reasonable to assume that there was already some demand, although construction with roof tiles and plaster was probably more unusual among the local common people, compared with Okinawa Island with a more urban culture. After the lifting of the restrictions on the use of roof tiles, demand for plaster increased, as in other parts of Okinawa. However, until the beginning of the twentieth century, the plasterers on Ishigaki Island were from Okinawa

Island and stayed there only while they worked. Then, gradually, some of them came to reside on Ishigaki Island and brought their families (Editorial Committee on History of Ishigaki City 1995: 830). Interestingly, on the island, even after World War II, traditional construction styles and techniques were not replaced as quickly as on Okinawa Island (Higa 2006: 78). Thus, the "construction rush," using roof tiles and plaster, lasted until about 1960, and it seems that quite a few local-born plasterers emerged during this time. However, in the 1960s, some destructive typhoons hit Ishigaki Island, prompting the diffusion of new materials and new building styles. It was 1967 when the last tile factory closed, and the number of plasterers also declined rapidly (Editorial Committee on History of Ishigaki City 1995: 830). In 1989, Mr. MN, a plasterer whom I describe later, opened a new tile factory, partly to revive the traditional construction culture, but closed it down after only a few years.

5.3 Stories from Plasterers on Ishigaki Island

Presently, on Ishigaki Island no craftsman is producing plaster from raw materials. Many of the more elderly people remember that there were quite a few kilns for plaster, but they are all now gone, without a trace. At least two plasterers—Mr. TU and Mr. MN, who are retired—and one widow of a plasterer, Mr. AB's then wife, survive. I interviewed all three regarding a traditional plasterer's work.

It is rather difficult to clarify the types of work plasterers have undertaken in the Okinawan context. There are at least four independent skills related to plastering: (1) acquiring corals as raw materials, (2) burning corals to produce quicklime, (3) processing quicklime into plaster, and (4) applying plaster to roofing with tiles. Strictly, those engaged, at least, in (4) are "plasterers," but particularly on the remote islands, such as Ishigaki Island, craftsmen tended to have to cover multiple skills. In fact, Mr. TU performed (2), (3), and (4), and Mr. MN and Mr. AB practiced all four skills. Moreover, Mr. MN had even produced roof tiles, as I mentioned. I interviewed them specifically regarding the process from (1) to (3).

Table 5.2 indicates the basic social attributes of the three plasterers. All started as plasterers before World War II (Table 5.2). As mentioned, on this island, the "construction rush" continued from the 1940s until around 1960, and, during this era, two different types of plasterers were active: "plasterers for generations," like Mr. TU, and "new entry plasterers," like Mr. MN and Mr. AB, who had started in their generation. In detail, Mr. TU's grandfather and father were plasterers in Shuri, Okinawa Island, and first came to the island to take jobs by themselves. After they decided to settle down on Ishigaki Island, they then brought their families over from Shuri. Plastering was their family business, and Mr. TU started his career when he was young, as an assistant to his father. In contrast, Mr. MN and Mr. AB learned their skills from other plasterers working on Ishigaki Island. Interestingly, Mr. AB used to be engaged in trading plaster produced on Okinawa Island, perhaps using connections to his father's homeland, Shuri, before becoming a plasterer himself. Mr. MN had experience in a number of jobs when he was young and then started working at a roof tile factory to master various skills, including plastering, on Ishigaki Island. He had also worked on Okinawa Island as a plasterer before coming back to his homeland to be independent. It is interesting that Mr. TU referred to the "new entry plasterers" as "jacks of all trades" and made a remark that there had been no professional plasterer originally on the island; he was clearly proud of his family and genealogy as the "legitimate" plasterers who could trace their ancestry back to having lived in Shuri on Okinawa Island and worked for the Ryūkyū Kingdom.

Here, I will describe the process for producing plaster. The three plasterers primarily acquired fresh corals as raw materials (Fig. 5.3). There were basically two ways of gathering coral: self-gathering and purchases from local fishermen. Mr. AB and his wife, living in Shiraho, used to go to the nearby lagoon at low tide. They walked on the dry or shallow parts to reach the reef crest and gathered corals particularly on the fore reef side. His wife said they snapped the roots of corals with iron sticks and put them in straw baskets. When a basket was filled, they carried it to the lagoon and asked a fisherman to carry it to the beach using his boat. She said that many kinds of corals were processable but "flat ones" were better than "pointed ones" because of density. On the other hand, Mr. TU and Mr. MN basically purchased corals from certain local fishermen who gathered living corals in addition to fishing. Many of the full-time fishermen were originally from other islands, such as Miyako Islands or Itoman, located at the southern part of Okinawa Island. However, both Mr. TU and Mr. MN sometimes also picked up dead dry corals on the beach or reused those from old houses.

The plaster kilns were built by the plasterers themselves using clay, tiles, and blocks (Fig. 5.4). It seemed common to build them on the beaches, but Mr. MN's was located further inland. They were also on the outskirts of the town because of the smoke and smell from burning corals. The method used to produce plaster from corals was common among the three plasterers, although there were also some subtle differences. Once they acquired corals, they dried them if they were raw or wet. Then, they piled corals in their kilns carefully, so as not to block the numerous vents, and burned them with firewood or waste oil and tires later. Corals need to be burned for approximately "3 days and 2 nights" or

Table 5.2 Social attributes of the three plasterers

Name	AB	TU	MN
Year of birth	1911	1928	1933
Place of birth	Shiraho, Ishigaki Island	Shika, Ishigaki Island	Shika, Ishigaki Island
Parents	From Syuri, Okinawa Island	From Syuri, Okinawa Island	Father from Shimane, Honshu. Mother from Ishigaki Island.
Background	Trader of plaster in Shiraho → Plasterer in Shiraho	Plasterer in Shika since young	Various jobs → Plasterer in Shika → Plasterer in Naha → Plasterer in Shika
Type of plasterer	"new entry plasterer"	"plasterer for generations"	"new entry plasterer"

Fig. 5.3 Carts full of corals in Shika, Ishigaki Island, in 1961. © Makio Andō (Reprinted, with permission, from Yaeyama Album: Track of 20th Century, Vol. 2 Ishigaki City 2001: 179)

about 50 hours, so at least two people were involved in continuously providing fuel.

After they stopped fueling the kilns and the fires were extinguished, the kilns were allowed to cool naturally for a day, and then the burned corals were taken out. By that time, the corals should have converted into white quicklime without losing their original shapes. The corals were then pulverized and covered with chopped rice straw; water was added and mixed. At this time, care was necessary because of the heat released from the mixture as a result of the chemical reactions. Then, they pounded the mixture in a mortar with a pestle and with machines later. This used to be thought of as a woman's job. Mr. MN said approximately 500 bags of 20 kg plaster were produced at one time, but the

Fig. 5.4 Lime kilns owned by Mr. TU in Shika, Ishigaki Island, in 1965. © Hiroo Ōnaka (Reprinted, with permission, from Yaeyama Album: Track of 20th Century, Vol. 2 Ishigaki City 2001: 179)

volume depended on the size of the kilns and the amounts of corals.

When they applied plaster on roofing with tiles, they carried plaster to the site and mixed it with water and sand. Apart from using it themselves, they also sold quicklime or plaster to others. At the time, quicklime was also used in producing brown cane sugar.

5.4 Concluding Remarks

From this short report on plaster and plasterers on Ishigaki Island, I would like to emphasize three points.

First, I reconstructed the process of traditional plaster production, which has not been recorded fully, at a time when the number of traditional plasterers is decreasing. It is now clear that, at least on the islands of Okinawa, corals from the sea were once gathered quite intensively as raw materials for plaster and were irreplaceable natural resources in the local construction culture.

Second, I found that the supply and demand for plaster on Ishigaki Island had long been limited because of legal restrictions and the distance from Okinawa Island during the era of the Ryūkyū Kingdom. The restrictions were lifted when the kingdom was abolished, and the situation then started to change. Particularly, after World War II, plasterers flourished in response to the "construction rush," but in the 1960s, they declined because of the introduction of new materials and building styles. Both events happened within short periods. The enactment of the Sea Area Fisheries Adjustment Regulations in 1972 truly put an end to the use of corals. In addition, the social perspective on corals has shifted from being regarded as natural resources to being understood as objects of conservation since the 1970s, and, thus, the construction culture of corals has been almost forgotten among the present local inhabitants.

Finally, I discovered the unique and non-monolithic nature of plasterers as a group on the island. The island had been located on the periphery since the era of the Ryūkyū Kingdom. As a result, there are various inhabitants who have emigrated, or whose ancestors have emigrated, from other islands for different reasons in addition to the original inhabitants. The "plasterers for generations" from Shuri or Naha were the ones who introduced and gave root to this traditional skill on the island; then, the "new entry plasterers," with various backgrounds, joined them when demand for plaster was high during the "construction rush." This shift can be understood as a process of localization on the part of the plasterers.

It is unfortunate that today corals are not to be touched and used, but only to be seen and cared for, within the context of nature conservation. I doubt whether this recent trend will lead to the successful conservation and recovery of the local corals, because it lacks a local perspective, based on traditional practices. We need to respect the history of coral use, recognize the possibility of various concerns about corals, and examine layered cultural values in the context of local inhabitants' lives. I believe that local-centered sustainable projects are only possible by doing this.

References

Higa T (2006) A study on the cement roof tiles of Sakishima Islands, Miyako and Yaeyama [Sakishima (Miyako Yaeyama) ni okeru semento gawara no ichikousatsu]. J Nago Mus: Ajimā [Nago Hakubutsukan Kiyou: Ajimā] 13:77–95. (in Japanese)

Ishigaki City (1995) History of Ishigaki City, itemized discussions: folklore vol. 1 [Ishigaki shishi kakuronhen: Minzoku, Jyō]. Editorial Committee on History of Ishigaki City. (in Japanese)

Ishigaki City (1999) History of Ishigaki City, series 13: Yaeyama Islands Annual Records [Ishigaki shishi sousho 13: Yaeyamajima nenraiki]. Editorial Section of the General Affair Department, Ishigaki City. (in Japanese)

Ishigaki City (2001) Yaeyama album: track of 20th century, vol. 2 [Yaeyama shashinchō: 20 seiki no wadachi, Gekan]. (in Japanese)

Ishii R (2010) Archaeology of roof tiles of the Islands: stories of Ryūkyū and roof tiles [Shimagawara no koukogaku: Ryūkyū to kawara no monogatari]. Shintensha. (in Japanese)

Kobashigawa T, Mezaki S (1989) Coral sea of Shiraho, Ishigaki Island: surviving miracle coral reef, revised and enlarged version [Ishigakijima Shiraho sango no umi, Zouhoban: Nokosareta kiseki no sangoshō]. Kobunken. (in Japanese)

Kuniyoshi F (2004) Plaster as a molding material: introduction of the Okinawan plaster study [Zoukeisozai to shiteno shikkui ni tsuite: Okinawa shikkui kenkyu josetsu]. Bull Okinawa Prefectural Univ Arts [Okinawa Kenritsu Geijyutsu Daigaku Kiyou] 12:37–50. (in Japanese)

Kyūyō Kenkyūkai (1974) *Kyūyō* [Kyūyō]. Kadokawa Shoten Publishing. (in Japanese)

Miyagi F (1982(1972)) Yaeyama lifestyle records [Yaeyama seikatsushi]. Okinawa times. (in Japanese)

Miyara T (1930) Yaeyama vocabulary [Yaeyama goi]. Toyo bunko. (in Japanese)

Noike M (1990) Living in the coral sea: life and nature in Shiraho, Ishigaki Island [Sango no umi ni ikiru: Ishigakijima Shiraho no kurashi to shizen]. Rural culture association Japan. (in Japanese)

Okinawa Prefecture Website (in Japanese) http://www.pref.okinawa.jp/reiki/34790210014300000000/34790210014300000000/34790210014300000000.html. Available on 07 Jan 2014

Possible Articulations Between the Practices of Local Inhabitants and Academic Outcomes of Landscape History: Ecotourism on Ishigaki Island

6

Kentaro Shimoda

Abstract

This chapter discusses possible practical articulations between local or personal knowledge related to the exploitation of living resources and academic outcomes of landscape history, based on the study of "ecotourism" on Ishigaki Island, Japan. The narratives of certain ecotourism company officers born and raised on Ishigaki remind us that the importance of the "skills and knowledge required to observe nature," based on biological or ecological knowledge, is relative. I suggest that fieldwork and research results may function as (1) a medium to connect people and (2) a catalyst to evoke local inhabitants' memories. If a course can be developed that makes effective use of these people's memories in future landscape planning, in all likelihood, ecotourism, as well as coexistence with nature on Ishigaki, will improve.

Keywords

Ecotourism • Landscape history • Local knowledge • Outsider • Participatory action research • Ishigaki Island

6.1 Introduction

This chapter discusses possible practical articulations between local or personal knowledge related to the exploitation of living resources and academic outcomes of landscape history, based on the study of "ecotourism" on Ishigaki Island, Japan. I focused on "ecotourism" because I wanted to provide a concrete example for local inhabitants so they might recognize ways to achieve beneficial exploitation of natural resources. In this chapter, I define "ecotourism" as "the medium that creates an awareness of nature" (Kikuchi 1999:145). I do not consider "ecotourism" a rigid concept. I also discuss the kinetic relationship that exists between local inhabitants and outside frames such as academic outcomes.

This chapter is organized into five sections. Section 6.2 provides an overview of "ecotourism" on Ishigaki Island. Sections 6.3 and 6.4 describe problems that occurred when academic outcomes were utilized by an "ecotourism company" on the island. The majority of my research was based on interviews and participant observation. The last section describes possible practical articulations between local or personal knowledge and academic outcomes of landscape history. It also examines the activities of the B02 research group that conducted fieldwork and outreach on Ishigaki Island.

6.2 "Ecotourism" on Ishigaki Island

Okinawa became a protected object when it was "discovered" as a tourist attraction (Yanaka 2007). However, scholars began to describe the coral reef of Okinawa based on a framework in which "development" and "protection" conflicted. As a result, chaotic development that caused the

K. Shimoda (✉)
Faculty of Core Research, Ochanomizu University, 2-1-1 Otsuka, Tokyo, Bunkyo, Japan
e-mail: basic_channel7@yahoo.co.jp

H. Kayanne (ed.), *Coral Reef Science*, Coral Reefs of the World 5, DOI 10.1007/978-4-431-54364-0_6

destruction of the coral reef ceased. However, mass tourism became a new problem because it exerted negative influences on the natural environment. Many advanced economies have faced similar situations. The United Nations declared 2002 as the International Year of Ecotourism. This marked a turning point because 10 years had passed since the "Earth Summit" was held in Brazil in 1992. The United Nations also described "ecotourism" as an effective action that might be used to address global environmental issues.

The tourism industry in Okinawa was developed based on the Okinawa Promotion Development Plan and the Okinawa Promotion and Development Special Treatment Act. The establishment of both laws was triggered by Okinawa's reversion to Japanese protection. They also resulted in the creation of the Okinawa Tourism Development Basic Plan. In addition, during the bubble economy, the Law for Development of Comprehensive Resort Areas (the "Resort Law") was enacted. With respect to the number of tourists who visit the remote island of Okinawa, the increasing rates of tourists who visit Ishigaki Island are remarkable. Tourism increased after 1997 when Ishigaki City announced its intention to develop tourism. Approximately 730,000 people visited in 2009 (Okinawa prefecture 2012). Although tourism has become an important economic industry for the island, the environmental burdens that developed from increases in tourism have raised many questions (Ishigaki City 2010).

The first ecotourism-related organization developed on the island, "Ishigaki Island Ecotourism Network" (EN), began operations in 2000. However, EN suspended its activities after several years. In 2005, the "Ishigaki Leisure on the Coast Safety Conference" (LCSC) was initiated. One year later, LCSC developed "ecotourism guidelines." The "Ishigaki Island Western Guide Association" (WGA) was begun in 2010. Its purposes were to formulate guidelines for ecotourism and to develop rules for ecotourism-related organizations (*Yaeyama Mainichi Shimbun*, June 9, 2010). In addition, the "Ishigaki Island Ecotourism Association" (EA) began operations in 2011. These recent activities suggest that the concept known as "ecotourism" was not uniformly understood on the island because only a few individuals belong to each of these groups, with the exception of several members of EN, which has suspended its activities. I believe each group attempts to develop tourism based on environmental conservation in its own manner.

In this paper, I examine tourism companies that provide nature-related experiences. These companies have names that (1) include the term "ecotourism," (2) advertise "ecotourism" on their home pages, or (3) advertise in magazines as "ecotourism companies" (hereafter, "company"). Based on my examinations of home pages, magazines, and interviews, I discovered that 27 companies located on Ishigaki Island conform to this standard. I also examined the hometowns of the representative officers of each company. I learned that, of the 27 representatives examined, 5 were born and raised on Ishigaki Island. Sixteen representatives had settled on the island. The remaining six individuals' backgrounds were unavailable. In other words, more than half of the 27 companies' representative officers were migrants.

6.3 "Islander" and "Outsider"

Between 2010 and 2012, I interviewed seven representative officers of companies that operate on Ishigaki Island. I also interviewed eight concerned individuals, including local administrators and conservationists. I performed these interviews in Japanese. Hence, I translated the following narratives.

My informants, who were settlers on Ishigaki Island, tended to have strong attachments to the Island's natural environment, as well as strong awareness of environmental conservation. For example, Mr. A (a man in his 50s), who began his company in the 1990s, stated: "I understood that the potential of Ishigaki Island was great. The sea that surrounds this Island is better than the sea that surrounds Iriomote Island" (March 12, 2012). Mr. B (a man in his 50s) was born on the main island of Okinawa. He settled on Ishigaki Island in 1981. He stated, "I wanted to live on Ishigaki Island because I was attracted to the nature environment here. The Island has the sea, the mountain, and the river" (May 19, 2011). Most of my informants participated as park volunteers for the Japanese Ministry of the Environment.[1] They regularly participated in voluntary cleaning activities. As can be seen in the responses of Mr. A and Mr. B ("The sea that surrounds this Island is better than the sea that surrounds Iriomote Island" and "The Island has the sea, the mountain, and the river"), most respondents appeared to focus on the importance of "nature."

I observed an additional characteristic of company officers who migrated to the island: They frequently ascribed differences in awareness between themselves and "islanders" to the fact that they were "outsiders." Several participants mentioned the image of the "islander" who ignores the destruction of the environment and the coral reef. For example, Mr. E (a man in his 20s who came from Tokyo) stated:

> Some fishermen believe there should not be a lot of coral reef. They offer various reasons. For example, a coral reef can

[1] The Ministry of the Environment in Japan registers people who voluntarily cooperate with various activities, such as interpretation and the observation of nature in the national park or the provision of cleaning as "park volunteers," to disseminate and promote knowledge of nature conservation.

> obstruct ships, destroy fishing nets, obstruct walking in the lagoon, and so on. Children are taught in elementary school that coral is important. Their fathers and grandfathers do not worry about the erosion of red soil (that exerts negative effects on coral reefs). I am hoping to find a balance between these two sets of conflicting ideas. (May 20, 2011)

In addition, when I asked Mr. C (a man in his 40s who came from Tokyo) whether the elderly "islanders" had taught him wisdom and traditions related to living on the island, he responded:

> We are not very good at that. If anything, elderly islanders have destroyed nature. Because I serve as an officer of the community center and participate in the PTA, I have opportunities to speak with them. However, all they said was that they had eaten fish that floated on the river because pesticide had flowed into the river. This type of story isn't suitable for ecotourism. (August 26, 2010)

Alternatively, an "islander" also spoke about differences in consciousness that exist between "insiders" and "outsiders." Although one man who worked in ecotourism admitted that he recognized the intense efforts of "outsiders" to conserve nature, he also stated, "It seems like outsiders have a different way of thinking from ours" (August 20, 2010). These narratives suggest that individuals who work in ecotourism are conscious of the dualism that exists between "islanders" and "outsiders."

6.4 The Skills and Knowledge Required to Work with Nature

I participated in "ecotours" conducted by five companies. One of these tours traveled around Ishigaki Island in 1 day. It visited the Hirakubo Cape, an observation platform located on the Tamatori Cape, the Gulf of Kabira, a famous coral reef observation point, the colony of Yaeyama palms located at Yonehara, the mangrove swamps located near the Fukido River, the Nagura River located in the western part of the island, and so on. In comparison with the first tour, the other four "ecotours" provided nature observation, snorkeling, and canoeing in narrow inlets. I made comprehensive notes of each guide's descriptions of nature and culture.

Table 6.1 Names provided by guides during one eco-company's tour

Creature	Plant	Topography	Folklore
Coral	Dragon fruit	**Reef**	***Kuchi***
Pheasant	Punai nut	**Lagoon**	***Okinawa-guchi***
White-breasted water hen	Chinese taro	Hilly area	***Daruma-guchi***
Peacock pansy	Screw pine	Plain area	***Yasura-guchi***
Coconut crab	Jack in the box		*Yarabu*
The common tiger	*Oobainubiwa* (*Ficus septica*)		*Muchi*
Malayan eggfly	Sea mango		*On*
Ghost crab	Fern palm		
Iso-hamaguri (*Atactodea striata*)	Fan flower		
Green turtle	Heliotrope tree		
The common rose	Indian beech *Pongamia* tree		
Fiddler crab (*Uca lactea* lacteal)	Red mangrove		
Fiddler crab (*Uca vocans*)	Seaside morning glory		
Barred mudskipper	*Otaniwatari* (*Asplenium antiquum*)		
Ryukyu tree lizard	*Komino-kurotugu* (*Arenga engleri*)		
Sakishima-tokage (*Scincella boettgeri*)	*Giran-inubiwa* (*Ficus variegata* var. *garciae*)		
Soldier crab	*Yaeyama-yashi* (*Satakentia liukiuensis*)		
Sand bubbler crab	Gray mangrove		
Red fiddler crab	Black mangrove		
Mud creeper	Akoh tree		
Shirena-shijimi (*Geloina coaxans*)	Chinese banyan		
Arasujikemangai (*Gafrarium tumidum*)	Sea almond		
Ryukyu-osagani (*Macrophthalmus convexus*)			
Ichijiku-kobachi (Agaonidae)			

Table 6.1 shows four categories ("creature," "plant," "topography," and "folklore"). It lists the names provided by tour guides during one eco-company's tour. I listed the names of creatures, plants, and topography provided by the guides in either standard Japanese or English. I listed those names under "folklore" in situations in which the names of creatures, plants, and topography were described in an indigenous language used on Ishigaki Island. Terms printed in boldface are names related to the coral reef. Table 6.1 provides a limited number of names related to the coral reef. This characteristic was common to all five companies. In addition, in Table 6.1, it is apparent that, although many names were provided for "creatures" and "plants," only a few names were provided for "topography" and "folklore." A comparison of the number of names provided during each of the five companies' tours in which I participated reveals that many names were provided for "creatures" and "plants" (Fig. 6.1). However, only a few names were provided for "topography" and "folklore."

Although I realize that the information I discuss was solely provided on five company tours, I noticed that, during those tours of Ishigaki Island, (1) the term "coral reef" was rarely

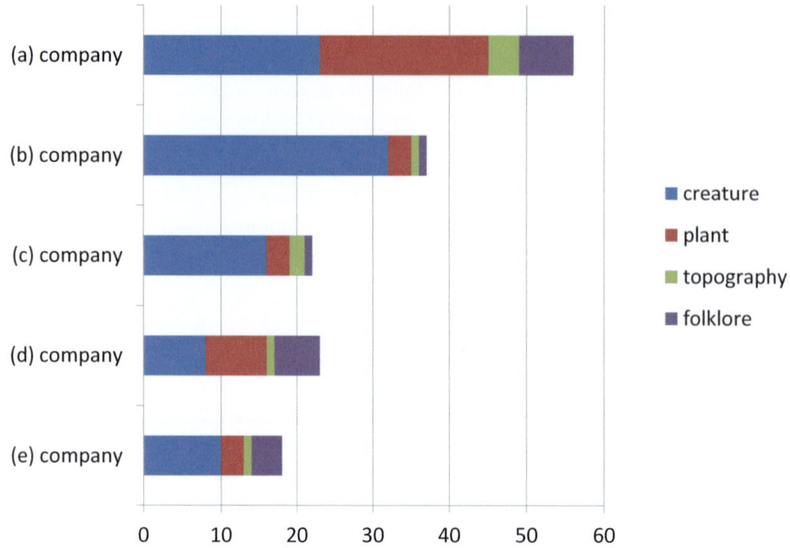

Fig. 6.1 Comparisons of a number of names provided during five company tours

used, (2) most of the tours included interpretations of biological or ecological knowledge related to creatures and plants, and (3) only a limited number of examples of indigenous knowledge were provided. These observations suggest that the skills required to interpret nature (i.e., the "skills and knowledge to observe nature") are widely used in ecotourism.

On the other hand, I discovered a contrasting feature of ecotourism in the guidebook to Iriomote Island (*Yamana Kara Suna Pitu*). This book is remarkable because it draws on the "skills and knowledge required to work with nature" that islanders and their ancestors cultivated in everyday life.[2] The book was written by an "islander," and its title translates to mean *The Mountain*, *The River*, *The Sea*, *The Human Being*. The book describes nature, creatures, plants, locations, and festivals found on Iriomote Island. It provides their names in indigenous languages. It also describes indigenous wisdom and skills related to fishing and the use of plants (Ankei et al. 1994). In contrast, as mentioned above, "skills and knowledge required to work with nature" were not consistently used in Ishigaki's ecotourism although "skills and knowledge required to observe nature" were emphasized. This may explain why only a limited number of "islanders" work as company officers. It might also explain the existence of the dualist consciousness that creates "islanders" and "outsiders."

An examination of the narrative of a company officer who was born and raised on Ishigaki Island can help us understand the concepts behind "skills and knowledge required to work with nature." Mr. F (a man in his 50s) is a fisherman who sometimes conducts tours in his fishing boat. He stated:

> I don't understand the way of thinking that states that the management of our resources equals environmental conservation. I acknowledge that the restoration of coral reefs can lead to more fish. However, I feel differently. (August 25, 2010)

He emphasized the "senses of fishermen" and stated that resource management was important to him. He fished with a fixed shore net placed near Nagura Bay each year from October to June. He stated:

> The sea around Nagura Bay is good. There are a lot of traces that might have been coral reefs in the old days. The thickness of the bar-shaped stony coral that lies beneath the seabed at around a depth of one meter is up to twice the size of a thumb. The reproductive life cycle of coral reefs in Nagura Bay has shortened in the last two to three decades. (August 24, 2011)

Mr. F also engages in fishing with basket nets each year from July to November. He stated:

> The coral reef will not die even if I break it in the sea. I use living coral to camouflage the basket nets. I break the coral reef and place the coral pieces around the basket nets. In many cases, the coral reef regenerates during a one-to two-year cycle. However, the coral reef in the neighboring sea areas of Kohama Island doesn't regenerate. (August 24, 2011)

Mr. G (a man in his 40s) offers canoe tours in his hometown. Mr. G was forced to move his office because of the construction of a resort condominium. Until 2011, he had located his office on an avenue lined with the huge *poon* trees planted by the ancestors of his village. These trees were saved from destruction when a plan to expand the national road was developed because officers from the village community center, including Mr. G's father, requested that

[2] This comparison of "skills and knowledge required to observe nature" and "skills and knowledge required to work with nature" is provided in Matsumura (2001), who analyzed ecotourism on Iriomote Island. In addition, in his analysis of the types of coexistence that occurred between human beings and coral reefs, Kakuma (2006) referred to a way of thinking that prioritized the ecosystem and biological diversity as "Western style." He referred to a way of thinking that prioritized the various types of connections that occur between human beings and coral reefs as "Asian-Pacific style."

Fig. 6.2 Avenue lined with huge *poon* trees (Photo by K. Shimoda, 2011)

Ishigaki City save the trees. Mr. G stated, "I wanted to set up my office near the avenue lined with the huge *poon* trees so that guests could view these trees" (May 20, 2011). According to Mr. G, these trees appear in many stories that involve his family and ancestors (Fig. 6.2).

Mr. G did not focus on preservation of the ecosystem. Rather, similar to Mr. F, Mr. G emphasized ways to work with nature. He stated:

> I can do something because I was born and raised on this Island. I can describe the scenery I saw in my childhood and I can describe the changes that have occurred. "Preservation" is impossible. I have watched the changes in my hometown and I want to preserve its memory for the next generation.... I want to acquire the most profound knowledge related to my home town. Where is the mangrove's pain? How has the forest located near the coast changed? How have creatures' behaviors changed? The environment is too big for me to understand. (May 20, 2011)

On March 13, 2012, Mr. G stated:

> I think it is important to feel the burdens of the environment. I want to increase the value of ecotourism. When I guided young girls during tours and spoke about the many creatures that live in the tidelands, at times, they stood still because they were afraid of stamping on creatures such as crabs. I think that is slightly incorrect. I spoke to them about how we walked on tidelands during our daily lives. I can never say that walking on tidelands and stamping on creatures is wrong. (March 13, 2012)

The ecotourism guidebook to Iriomote and the narratives of company officers who were born and raised on Ishigaki provide additional details related to "skills and knowledge required to work with nature." They also remind us that the importance of the "skills and knowledge required to observe nature," based on biological or ecological knowledge, is relative. In the following section, I discuss possible practical articulations between local or personal knowledge and academic outcomes of landscape history.

6.5 Possible Articulations Between Local Inhabitants' Practices and Academic Outcomes of Landscape History

If we consider ecotourism as "the medium to create new consciousness of nature," then a researcher may become an agent who exerts effects on ecotourism because the researcher will rely on scientific knowledge to discover nature. The research may play an important role in the evaluation of natural resources.

During my interviews, I observed that company officers focused significant attention on academic outcomes. For example, on March 12, 2012, Mr. D (a man in his 40s) stated, "I read research papers available on the Internet." In addition, another company officer showed me a bundle of research papers he maintained in his home. Mr. H (a man in his 50s) belongs to EA. He stated that he considered himself an intermediator whose role was to articulate academic outcomes for the common people. He attached importance to conveying these outcomes in ways that would be easy for all people to understand (August 22, 2011).

The B02 research group conducted lectures in Ishigaki on 4 days in August 2010 and August 2011 to describe the results of our research. My four informants attended these lectures. The B02 research group hoped to construct a local history focused on the interrelatedness of humans and nature. We investigated the micro-atolls that dot Nagura Bay with Mr. B in March 2012. In this section, I discuss the impact of academic results on the practice of "ecotourism," as well as on "local inhabitants," based on these outreach activities.

A narrative provided by Mr. H demonstrates how ecotourism guides can translate academic outcomes into "topics" that might interest tourists. Mr. D, who attended a lecture provided by the B02 research group, stated:

> An excellent guide has a wealth of knowledge. However, he/she also makes the tour amusing for the customers. We need to know many topics to amuse the customers. The ordinary tourists always ask me whether nuts or shellfish they see are edible. Not all of the customers who visit Ishigaki have had experiences with nature. Most hope to free themselves temporarily from their ordinary daily lives.... I studied quite hard after I decided to start an ecotourism company.... I am thankful for your lecture. (March 12, 2012)

Although each guide presents unique topics, all of my informants create their topics based on "knowledge" (e.g., biological, ecological, historical, topographical, folkloric knowledge, and so on).

One of the goals of the B02 research group was the creation of a scenario that described the environmental history of Nagura and focused on the interrelatedness of humans and nature. In addition to attending our lecture, Mr. B also assisted in our investigations of the micro-atolls. With respect to our research results, he stated:

> I explained that the bedrock in Nagura rose up about 3,000 years ago. However, the numeric data related to the micro-atolls is very important. I already knew about the existence of micro-atolls in Nagura. Yet, the investigations performed by the B02 research group explain the importance of the micro-atolls' existence in an entertaining way. (May 19, 2011)
>
> During the lecture, I learned that Nagura's environment was not formed by sea level changes. Rather, it was formed by erosion. Although my estimate was wrong, this new insight was interesting.... I share my knowledge, as well as my experience when I conduct ecotours.... The fact that I participated in this kind of investigation and acquired this knowledge adds to my story. The fact that I attended this lecture will also add to my story. (March 14, 2012)

These narratives suggest that Mr. B recognized that the micro-atolls in Nagura were valuable resources. He also attempted to incorporate stories based on his experiences during our investigation and lectures into his tours. The case of Mr. B suggests that the research results, as well as the investigation of the local environmental history, enlarged the parameters of what might be considered "resources" in the natural environment.

Finally, I would like to discuss possibilities offered by research-related outreach activities. In this paper, I described the struggles that can occur between "islanders" and "outsiders." I also discussed the differences that might exist between "skills and knowledge required to observe nature" and "skills and knowledge required to work with nature." These differences might be caused by dissonances that occur among people involved in ecotourism in Ishigaki. However, the history and reality of Ishigaki demonstrate the risks involved in our conceptions of this type of dichotomy. A variety of individuals began to immigrate (e.g., farmers during the Ryukyu Kingdom era and Itoman fishermen prior to the war, free immigrants who left Miyako Islands) (Miki 2010). Currently, "participatory action research" (PAR) has been the focus of increasing attention.

PAR is a social practice that involves collaborative learning realized by groups of people who join to solve problems and promote dialogues among the concerned people. The PAR process involves a spiral of self-reflective cycles such as planning a change, acting, observing the process and consequences of the change, reflecting on these processes and consequences, and replanning (Kemmis and McTaggart 2005). During "participant observation," the traditional investigation method used in the social sciences, the researcher tends to avoid participation because it might change the object. The researcher places high value on comprehension of the object as it truly exists. In contrast, PAR researchers presuppose that their participation will cause direct changes on the object.

Figure 6.3 shows the results of a questionnaire we distributed to learn the occupations of attendees to lectures provided by the B02 research group between 2010 and 2011. People aged between their teens and their 80s who performed a variety of occupations (e.g., teachers, students, researchers, office workers, public servants, farmers, local historians, dyers, and ecotourism company officers) participated in these lectures. I believe the lectures' theme, "local history," was one of several reasons why this variety of people attended (cf. Marcucci 2000). Ample time was provided to encourage pleasant conversations during circles that occurred after lectures were held in Nagura Community Center on August 19, 2011. Many people enjoyed sharing their memories of older ways of life with other participants (Fig. 6.4). In this case, fieldwork and research results may function as (1) a medium to connect people and (2) a catalyst to evoke local inhabitants' memories. If a course can be

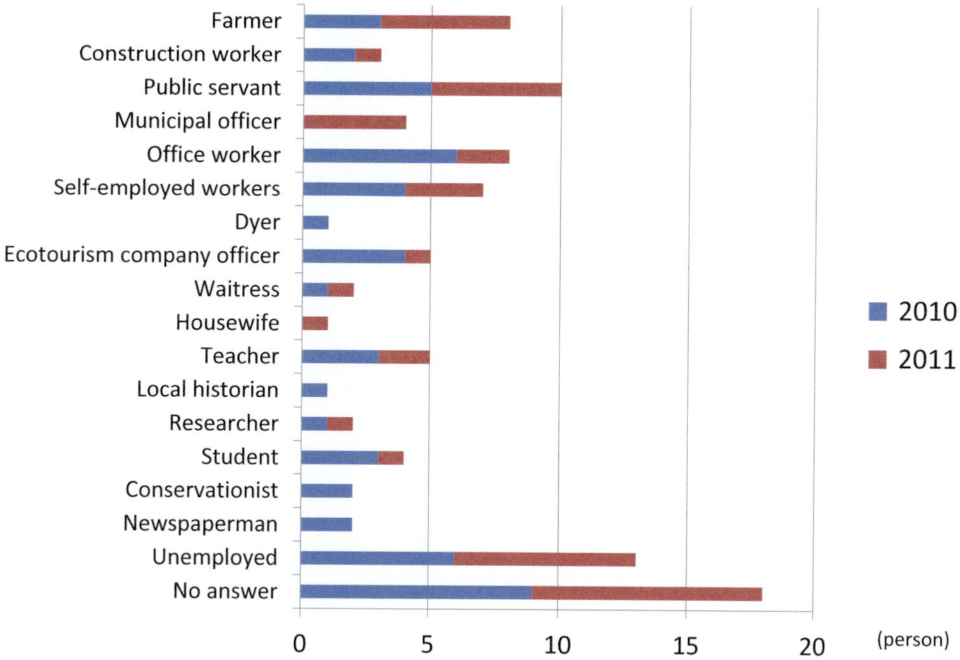

Fig. 6.3 The occupations of visitors to our lectures

Fig. 6.4 Pleasant talks held after the lecture (Photo by K. Shimoda, 2011)

developed that makes effective use of these people's memories in future landscape planning, in all likelihood, ecotourism, as well as coexistence with nature on Ishigaki, will improve.

Acknowledgments Many people contributed to the completion of this article. I received generous support from Ishigaki City Board of Education, Ishigaki City Library, Nagura Community Center, and Yaeyama Museum. I most sincerely thank all the interviewees who generously shared their practices and memories with me.

References

Ankei Y, Ishigaki A, Ishigaki K, Itani G, Oba N, Sakaguchi N, Satake K (1994) Yamana Kara Suna Pitu: Iriomote Island ecotourism guidebook. Iriomotejima Ecotourism Kyokai, Okinawa (In Japanese)

Ishigaki City (2010) *Ishigaki shi kanko kihon keikaku* (In Japanese)

Kakuma S (2006) The challenges and methods in MPA, coral reef conservation, and ecotourism in tropic and sub-tropic regions: a case study of Zamami in Okinawa and the Asia-Pacific Islands. In: Shinzaki S (ed) A study of a sustainable development model for small island communities near Naha, Okinawa confronting depopulation and aging (FY2003–2005 final research report of grant-in-aid for scientific research (B), pp 101–119

Kemmis S, McTaggart R (2005) Participatory action research: communicative action and the public sphere. In: Denzin N, Lincoln Y (eds) Handbook of qualitative research, 3rd edn. Sage Publications, Thousand Oaks, pp 559–604

Kikuchi N (1999) Toward a perspective of environmental sociology on eco-tourism: reconsidering "Inhabitants" and "Nature". J Environ Sociol 5:136–151

Marcucci DJ (2000) Landscape history as a planning tool. Landsc Urban Plan 49:67–81

Matsumura M (2001) General survey of tourism studies of Yaeyama Islands: dynamic relations of people in Taketomi, Iriomote, and Kohama Islands with nature. Environ Dev Cult Asia-Pac Soc 2:140–202

Miki T (2010) Yaeyama gashukoku no keifu. Nanzansha, Okinawa (In Japanese)

Okinawa Prefecture (2012) *Kanko yoran heisei 23 nendo ban.* (In Japanese)

Yanaka S (2007) Shakai kankei no naka no sigen: Watarama kaiki sango sho o megutte. In: Matsui T (ed) Shizen no sigenka. Kobundo, Tokyo, pp 83–119 (In Japanese)

Response of Coral Reefs to Global Warming

7

Hajime Kayanne

Abstract

The atmospheric CO_2 crossed over 400 ppm and we must prepare for a "+2 °C world" during this century. Coral reefs are directly related to each scenario of global warming: increase in CO_2 results in ocean acidification and suppresses calcification, rise in sea surface temperature leads to severe bleaching, and sea level rise causes submergence of coral reefs and atoll islands. They are the most sensitive ecosystem and act as an early warning system to examine and predict response of ecosystem to the global warming. Records of bleaching events and SST for the last 17 years in the northwest Pacific show that 2 °C SST rise would induce severe bleaching of coral reefs. Reduction of ocean surface water pH by 0.3 would cause shift from hard coral to non-calcifying macroalgae or soft coral populations in coral reefs. Submergence of coral reefs by sea level rise of up to one meter results in a loss of their breakwater function and in atoll islands submergence of national land. "+2 °C world" is the threshold to maintain coral reefs. Factors of global warming and responses of coral reefs are coupled to form feedback loops, which enhance or stabilize the changes within a system.

Keywords

Coral reef • Global warming • Ocean acidification • Sea level rise

7.1 Introduction: Coral Reefs in "+2 °C World"

In June 2013, atmospheric CO_2 concentration crossed over 400 ppm as a daily average for the first time after a steady rise during the monitoring at Mauna Loa since 1958 (Monastersky 2013). This level is the highest even for the last a few million years. Before the Industrial Revolution, it never reached 300 ppm, and this increase of CO_2 has induced +0.85 °C global average temperature increase, +0.19 m sea level rise, and 0.1 unit pH decrease through the twentieth century as reviewed by the latest Fifth Assessment Report of Intergovernmental Panel on Climate Change (IPCC AR5: http://www.ipcc.ch/).

For the future projection, AR5 adopted four Representative Concentration Pathways (RCPs) with prescribed CO_2 concentrations reaching 421 ppm (RCP2.6: the digit indicates total radiative forcing in year 2100 relative to year 1750 in Wm^{-2}), 538 ppm (RCP4.5), 670 ppm (RCP6.0), and 936 ppm (RCP8.5) (Moss et al. 2010). RCP2.6 scenario requires that we tackle to reduce CO_2 emission at once, and RCP8.5 scenario represents that we continue to rely on fossil fuels for our energy. If we do not adopt reduction of CO_2 emission at once (RCP2.6), global surface temperature rise at the end of the twenty-first century is likely to exceed 1.5 °C relative to 1850–1900 (Fig. 7.1a). Most researchers studying global change are afraid that

The original version of this chapter was revised. An erratum to this chapter can be found at DOI 10.1007/978-4-431-54364-0_9.

H. Kayanne (✉)
Department of Earth and Planetary Science, Graduate School of Science, The University of Tokyo, Tokyo, Japan
e-mail: kayanne@eps.s.u-tokyo.ac.jp

H. Kayanne (ed.), *Coral Reef Science*, Coral Reefs of the World 5, DOI 10.1007/978-4-431-54364-0_7

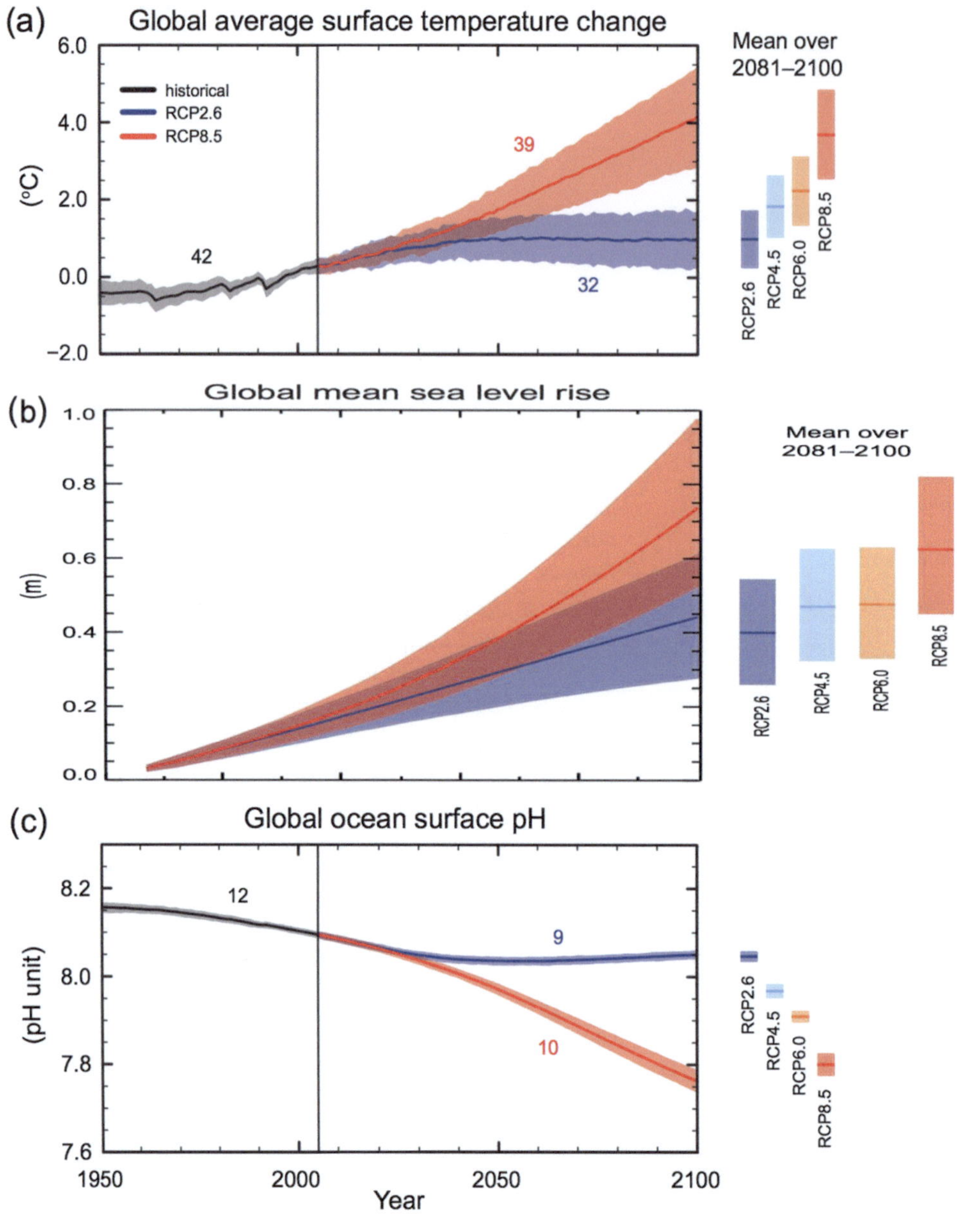

Fig. 7.1 Simulated time series from 1950 to 2100 relative to 1986–2005 for (**a**) change in global annual mean surface temperature, (**b**) global mean sea level rise, and (**c**) global mean ocean surface pH (AR5 of IPCC). Projections and uncertainty (*shading*) are shown for scenarios RCP2.6 (*blue*) and RCP8.5 (*red*)

"+2 °C world" would be inevitable and that we may face to "+4 °C world" as an extreme case (RCP8.5). The rise of sea level to be a worst case of +1 m by the end of this century (Fig. 7.1b) and surface ocean pH would decrease by 0.1–0.4 (Fig. 7.1c) units accompanying with CO_2 increase known as ocean acidification.

Ecosystem on the globe adapts to the present climate, and human activities are tied to the present ecosystem and climate. So the global climate change may collapse the ecosystem and human activities, which may not keep up with the rapid change. However, the global change so far is still small as compared with natural variability to extract its effect apparently, and we just project the probable scenario of response of ecosystem to the future global changes. The worldwide coral reef bleaching event in 1997–1998 gathered the attention not only from coral reef scientists but also from researchers in general interested in global change. The bleaching, loss of symbiotic algae from the host corals resulting in their mass mortality, was induced by ocean surface thermal anomalies 1–4 °C higher than normal years. It was the first event showing ecosystem-scale response to global warming. Before that event, coral reef researchers themselves had regarded local human impact as the main stressor to coral reefs, and global climate change would become the serious issue some time during the twenty-first century. After the bleaching event, they changed the mind and have now regarded global as well as local environmental change would evenly bring serious impact to coral reefs (Wilkinson 2002).

Coral reefs are directly related not only to warming but also to the other factors of the global warming scenario (Fig. 7.2). Coral reefs are the largest site of calcium

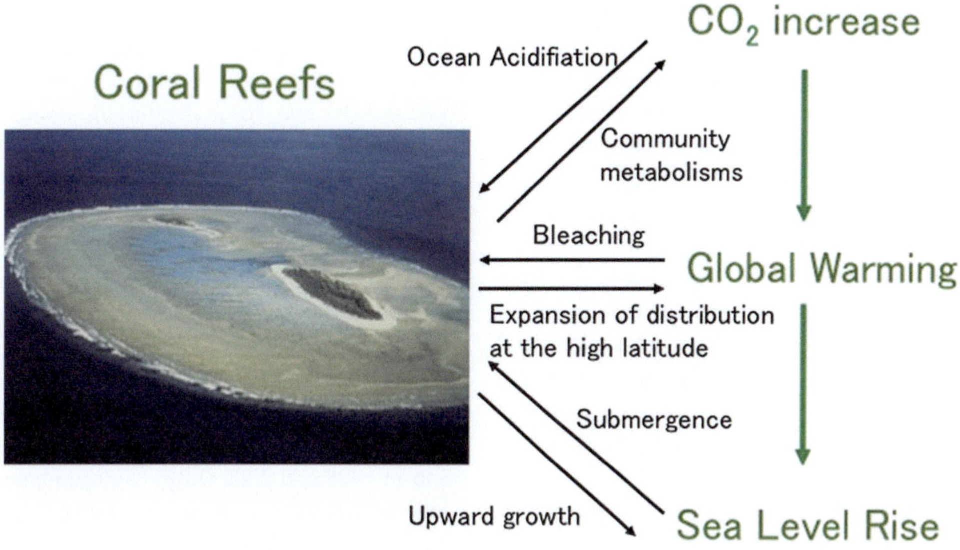

Fig. 7.2 Responses of coral reefs to global warming

carbonate production, which would be suppressed by the ocean acidification. The rise in sea surface temperature would cause more frequent and severe coral reef bleaching. Sea level rise would lead to submergence of coral reefs and loss of its function as a natural breakwater and habitat for the highest diversity of lives in the ocean.

Hoegh-Guldberg et al. (2007) inferred that density and diversity of corals are likely to decline under a global warming scenario of 450–500 ppm CO_2 and +2 °C temperature rise from preindustrial level, and macroalgae would be dominated over a few corals under a scenario of CO_2 >500 ppm and +3 °C thermal stress. The threshold CO_2 concentration of 450 ppm with +2 °C thermal stress for the decline of coral reef ecosystem has been reconfirmed by the later reviews on response of coral reefs to the global warming scenario (Veron et al. 2009; Hoegh-Guldberg and Bruno 2010). This threshold value corresponds to a moderate future scenario of the IPCC projection (RCP4.5 or 6.0) of "+2 °C world." Silverman et al. (2009) suggested a threshold of 550 ppm for coral reefs to erode rather than grow by acidification and surface ocean warming. These reviews predict the decline of coral reefs by the combined stresses of high SST and low pH. On the other hand, Frieler et al. (2013) claimed that only a thermal stress of +2 °C would be enough for two-thirds of world coral reefs to be subject to long-term degradation.

However, responses of coral reefs are not straightforward but will return feedback reactions against the change. Increase in CO_2 would also enhance dissolution of calcium carbonate sediments as well as photosynthesis, both of which act to reduce CO_2 concentration at least in coral reef water. Global warming would also expand the habitat of corals along the margin of their distribution. Coral reefs have a potential to catch up with rising sea level. Moreover, corals may adapt to the future ocean acidification and/or rise in SST.

We need to understand complicated responses of coral reefs to the global warming. By now, the signatures of the global warming are not so large (−0.1 pH unit ocean acidification, +0.85 °C temperature, and +0.19 m sea level rise by IPCC AR5). Coral reefs have already been affected by and responded to the global warming. Coral reefs are the most sensitive ecosystem and act as an early warning system to examine and predict response of ecosystem to the global warming. Analysis of the early signatures is effective to predict the future changes in coral reef response to the global warming.

7.2 Global Warming

7.2.1 Threshold Temperatures for Bleaching

Since the worldwide bleaching event in 1997–1998 (Wilkinson et al. 1999), fortunately no such global event has occurred. However, specific region of coral reefs suffered from regional scale higher than normal temperature resulting in regional-scale bleaching event. Field experience and laboratory experiment showed that thermal stress of 1–2 °C higher than the usual summer maximum can cause mass coral bleaching (Jokiel and Coles 1990). Small increases in SST (0.5–1.5 °C) over several weeks or large increase (3–4 °C) over a few days will lead to bleaching (Glynn 1993; Baker et al. 2008). Thus, the accumulation of thermal stress should determine the threshold for coral bleaching.

The National Oceanic and Atmospheric Administration (NOAA), USA, has monitored world sea surface temperature (SST) by satellite and released near-real-time anomalously high SST regions as coral bleaching HotSpots charts with 1° longitude × 1° latitude grid since 1997. The HotSpots anomalies are summed to estimate a degree heating week (DHW) to predict bleaching. DHW is the accumulation of the anomalies exceeding the maximum of the monthly mean SST for a given region over a rolling 12 weeks (Liu et al. 2003, 2006). Only anomalous values ≥1 °C are accumulated on an assumption that less than 1 °C SST anomaly is insufficient to cause visible stress on corals. Two DHW are equivalent to 2 weeks of anomalous temperature staying at 1 °C or 1 week of anomalous SST at 2 °C and so forth. Empirically, DHW values of 4.0 °C would cause bleaching, and as DHW values reached 8.0 °C, widespread bleaching is likely and some mortality is expected.

Kayanne (submitted) validated the DHW as threshold values for mass coral bleaching by the observed bleaching events in eight islands in the northwest Pacific: Kushimoto, Kochi, Amami Oshima, Okinawa Island, Ishigaki Island, Ogasawara Islands, Guam, and Palau (Fig. 7.3). In these localities coral bleaching events were observed in detail and recorded at least since 1998. Occurrence of the bleaching events is differed in year location by location and thus effective to test their correlation with temperature anomalies in these localities.

In Shiraho Reef, southeast coast of Ishigaki Island, a 15-year survey along fixed transect revealed time-series change in coral cover in response to thermal stresses (Harii et al. 2014). Coral population at Shiraho Reef decreased to a half after the 1998 bleaching, but recovered dramatically by 2000 (Kayanne et al. 2002). Coral coverage increased until 2003 after damage by 1998 bleaching, but then decreased by 2009 mostly due to a decline in branching *Montipora* (Fig. 7.4; Harii et al. 2014). The decline between 2003 and 2009 is explained by the bleaching event in 2007 (Dadhich et al. 2012) and mechanical destruction by large typhoons from 2004 to 2007. The bleaching event in 2007 was observed widely around Ishigaki Island; 60 % of acropoid corals died by the bleaching in Sekisei Lagoon, south of Ishigaki Island (Nojima and Okamoto 2008). In Sekisei Lagoon, moderate-scale bleaching was also reported in 2001 and 2003, but it was not observed in Shiraho Reef.

In Amami Oshima, island-scale bleaching was reported to occur in 1998 and 2001 (http://www.coremoc.go.jp/375). In Okinawa Island, two thermally induced bleaching events were recorded also in 1998 and 2001 (van Woesik et al. 2011). Bleaching events were also recorded in 1980, 1983, 1986, 1991, 1994, 1995, 1996, and 2003 in Okinawa (Nakano 2004). In Ogasawara Islands, widespread bleaching had not been observed since 1973 including 1998 (http://www.coremoc.go.jp), and it was the first event identified in September 2003 (Yoneyama et al. 2008). Guam was also escaped from bleaching in 1998. Since 1970, only two large-scale bleaching events have been recorded in 1994 and 1996, both of which did not appear to be associated with high SST (Paulay and Benayahu 1999; Porter et al. 2005). In Palau, severe bleaching occurred in 1998, but the coral cover recovered less than a decade later. However, in 2010 Palau experienced another thermal stress bleaching event (van Woesik et al. 2012).

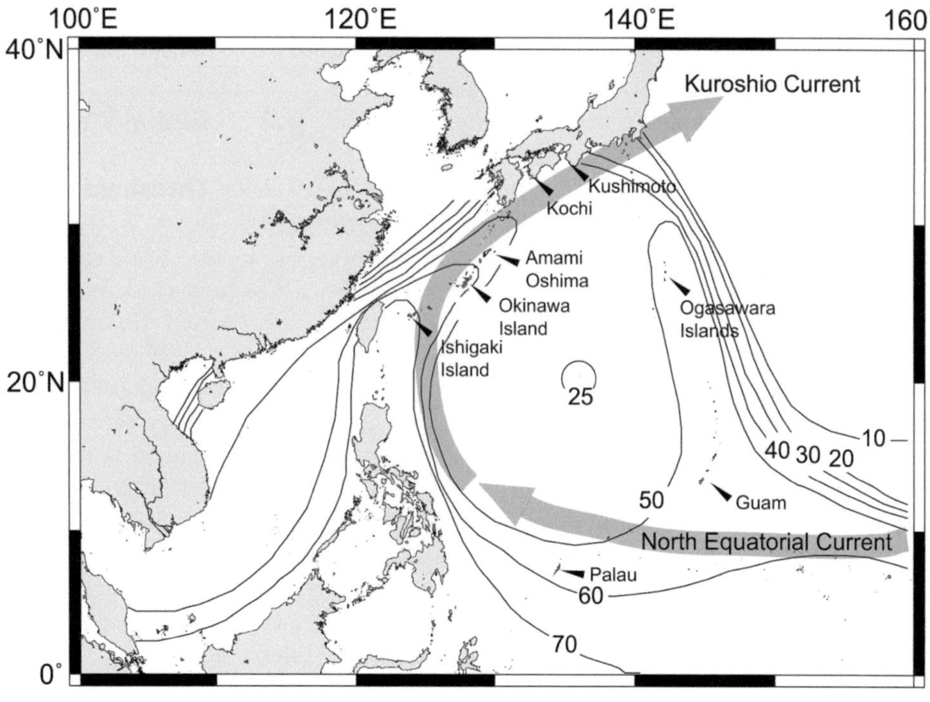

Fig. 7.3 Locality map for examining relation between bleaching events and SST anomalies. Isolines show number of coral genera

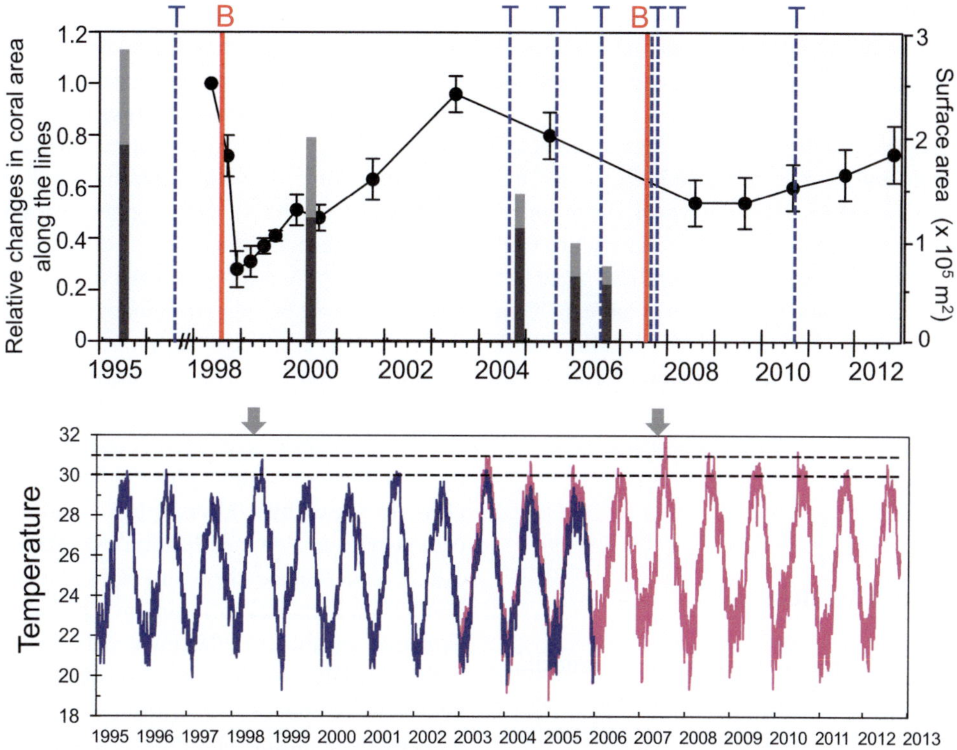

Fig. 7.4 (**a**) Time-series changes in total net area of six dominant coral genera/species along five transect lines from 1998 to 2012. The bars show surface area ($\times 10^3$ m^2) of coral coverage from 50 % to 100 % (*black bars*) and from 5 to 50 % (*gray bars*) in Shiraho Reef from 1995 to 2006 by aerial photographs. *B* bleaching events (*red lines*), *T* typhoons (*blue line*). (**b**) Changes in sea surface temperature from 1998 to 2012 at Ishigaki Port (*blue line*) and Shiraho Reef (*pink line*) (From Harii et al. (2014) with permission of Inter Research)

The observed bleaching events in the eight localities are examined retrospectively in relation to the satellite-based DHW (Kayanne, submitted). In general, DHW provides reasonable threshold for bleaching: DHW of >8 °C induced severe bleaching events. However, in some cases DHW less than 4 °C induced local bleaching events, which might be resulted from local heating event on shallow reef flats (Ishigaki Island in 2007 and several cases in Okinawa before 1996) or bleaching which was not resulted from thermal stresses (Guam in 1994 and 1996 and several cases in Okinawa before 1996).

DHW of >8 °C corresponds to 1-month averaged anomaly of >2 °C, which thermal environment should be easily attained by "+2 °C world" under the global warming. Therefore, unless we tackle with the global warming at once, coral reefs would suffer from severe bleaching every year, and we cannot sustain healthy reefs to the next century, unless they adapt to the thermal stress.

7.2.2 Poleward Range Expansion of Corals

Degradation of corals by thermally induced bleaching has obtained attention for tropical areas. On the other hand, more favorable condition would be expanded to the higher latitudes, and poleward shift in species distribution may occur. Average range shifts have been reported as 6.1 km/decade for terrestrial communities (Parmesan and Yohe 2003), induced by shift of isotherm on land with a velocity of 27.3 km/decade (Burrows et al. 2011). Meta-analysis of response of marine organisms to recent climate change revealed a mean rate of poleward expansion at the leading range edge of marine species was 72.0 ± 13.5 km/decade with faster rate for mobile pelagic organisms (Poloczanska et al. 2013).

Japan coast, along which coral fauna distributes accompanying with strong constraint of SST gradient (Veron and Minchin 1992; Sugihara et al. 2009), provides a unique site to monitor poleward shift of coastal fauna by warming. Yamano et al. (2011) analyzed 80 years of coral fauna records along Japan coast from 29°N to 35°N and found most major coral species showed poleward range expansion with a speed of 140 km/decade since the 1930s. Four major coral species categories (*Acropora hyacinthus, A. muricata, A. pruinosa, and A. solitaryensis*) showed poleward range expansions since the 1930s, whereas no species demonstrated southward range shrinkage. The expansion speed is higher than terrestrial or average marine fauna, which is explained by relatively larger increase in SST (+1.5 °C) along the mainland Japan coast and high dispersal potential of coral larvae delivered by the strong Kuroshio Current.

Model study also showed that coral habitats are projected to expand northward by several hundred kilometers by the end of this century (Yara et al. 2011). However, the northern expansion of coral distribution may be limited by southward shift of the isoline of carbonate saturation state of Ωaragonite = 3, as a limit for tropical coral habitat (Yara et al. 2012). The southern habitat will be suffering from more frequent bleaching, and the model simulation projected that coral habitat will disappear at the end of the twenty-first century under the scenario of "business as usual scenario" corresponding to RCP8.5.

7.3 Ocean Acidification

Anthropogenic increases in CO_2 cause ocean acidification, declining calcium carbonate saturation states ($\Omega = [Ca^{2+}] \times [CO_3^{2-}]/K_{sp}$), and reduced coral reef calcification (red arrows in Fig. 7.5). The effect of ocean acidification to marine calcifiers has been examined by increasing number of CO_2-enrichment experiments in laboratory, which are conducted only for a short time period (days to weeks) under fixed conditions. However, natural environment is changing its physical and chemical conditions daily, seasonally, and inter-annually. As the number of experiments grows, variation of responses to increasing CO_2 has become obvious (Doney et al. 2009). In a high CO_2 experiment conducted under near-natural environment, calcification rate of *Acropora digitifera* did not decrease with increase in CO_2 (Takahashi and Kurihara 2013). Moreover, organisms interact with each other and with physical environment. The relationship between calcification rate and CO_2 is most likely more variable than previously speculated.

Laboratory experiment cannot predict which organisms would replace the present dominant species (in the case of coral reefs, corals). Most projected community changes due to ocean acidification describe transitions from hard coral to non-calcifying macroalgal communities with an analogy from state shift of coral reefs by local stresses (Hoegh-Guldberg et al. 2007). Other organisms have received less attention, despite the biotic diversity of coral reef communities. Therefore, we need to shift our view not only from laboratory but also to actual fields.

In shallow coral reef flats, we observed a large diurnal variation for physical and chemical parameters. In Shiraho Reef, Ishigaki Island, Ryukyu Islands, during daytime dissolved inorganic carbon (DIC) decreases by 400 μmol kg^{-1} by calcification and photosynthesis, total alkalinity (TA) decreases by 100 μmol kg^{-1}, and CO_2 decreases by photosynthesis over calcification (Kayanne et al. 2005). The change in carbonate chemistry showed a large spatial variation by heterogeneity in community metabolisms and hydrodynamics on the reef flat (Watanabe et al. 2013). During nighttime, DIC increases by photosynthesis, and TA shows almost stable values, but in some cases TA shows a slight increase by nighttime dissolution, which was observed when the saturation state of aragonite (Ω_{arag}) in water of 2.5–3.8. It has been known that Mg-calcite with Mg concentration between 8 and 12 mol% as formed by foraminifera, sea urchin, and coralline algae dissolves more readily than aragonite or calcite (Plummer and Mackenzie 1974; Morse et al. 2006). The coral reef flat sediments contain much high Mg-calcite formed by foraminifera and calcareous algae.

Yamamoto et al. (2012) conducted a dissolution experiment and found that foraminifera tests and calcareous algae formed by high Mg-calcite start to dissolve at $\Omega_{arag} = 3.0$. Yamamoto et al. (2015) further explored the sediment-water interface on a calcareous sand in Shiraho Reef and found that Ω_{arag} in sediment pore water deeper than 5 mm had a constant value of 3.0, which equals the saturation

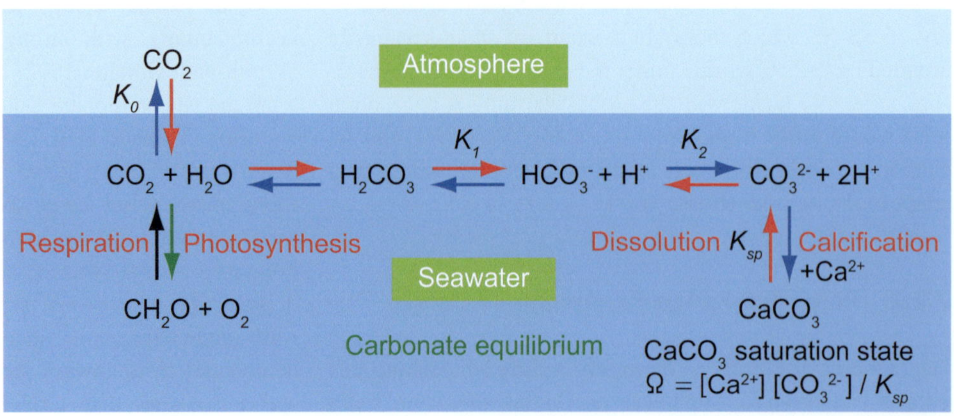

Fig. 7.5 Schematic diagram of ocean carbonate system

threshold of foraminifera. In the sediment pore water, CO_2 released by organic respiration is neutralized by inorganic Mg-calcite dissolution to keep constant saturation state. In the surface sediment, Ω_{arag} and TA had gradients, resulting in a constant TA flux into the water column, which indicates dissolution of Mg-calcite sediment. Dissolution may act as a buffer against increasing CO_2 within sandy areas of coral reefs.

To evaluate acidification impact on ecosystems, natural CO_2 seeps have now been attracted much interest. The first such site was reported from Ischia Island off Italy, Mediterranean Sea (Hall-Spencer et al. 2008). At this site, along gradients of normal pH (8.1–8.2) to lowered pH (mean 7.8–7.9, minimum 7.4–7.5), typical rocky shore communities with abundant calcareous organisms shifted to communities lacking scleractinian corals with significant reduction in sea urchin and coralline algal abundance and dominance of sea grass. The second one, the first from coral reefs, was reported from Milne Bay, Papua New Guinea (Fabricius et al. 2011). At this site, with declining pH from 8.1 to 7.8, coral diversity, recruitment, and the abundance of structural complex framework builders were reduced and the abundance of macroalgae increased, but pH did not affect the abundance of hard corals. Another in situ study of coral reef communities in Mexico also demonstrated that lowered pH derived from high-alkalinity groundwater decreased coral diversity (Crook et al. 2012). These sites suggested that coral community may shift to non-calcifying macroalgae or sea grass by lowered pH of 7.8 (=800 ppm CO_2).

Another CO_2 seep site from coral reefs was reported from Iwotorishima Island, an uninhabited volcanic island in the Ryukyu Islands, Japan (Inoue et al. 2013). Hard corals are restricted to non-acidified low-pCO_2 (225 ppm, pH 8.3) zones, dense populations of the soft coral *Sarcophyton elegans* dominate medium-pCO_2 (831 ppm, pH 7.8) zones, and both hard and soft corals are absent from the highest-pCO_2 (1465 ppm, pH 7.6) zones. Culture experiment confirmed the benefited effect of medium-level pCO_2 to *S. elegans* by enhancing photosynthesis while no effect on light calcification. These results suggest that reef communities may shift from hard coral to soft corals under pCO_2 of 550–970 ppm and challenge the survival of hard corals.

7.4 Sea level Rise

7.4.1 Response of Coral Reef Crest

Coral reefs typically have a shallow reef flat and a reef slope to its seaward. In many cases, a rise, called the reef crest, exists on the seaward edge of the reef flat, which acts as a natural breakwater to separate the reef flat from the open ocean. The reef flat and coastal area behind is protected from ocean waves and swells by the reef crest. Coral reef surfaces caught up with the sea level after its stabilization from 7000 to 4000 years ago with some interruption by environmental deterioration (Hamanaka et al. 2012). The first part of the reef to reach sea level was the reef crest with a vertical accumulation rate of 0.1–0.4 m/100 years (Kayanne 1992) or up to 0.5 m/100 years (Hongo 2012). Some fast-growing corals such as branching *Acropora* can add up to several meters/100 years (Montaggioni 2005), but they are sheltered or shallow water species deeper than 3–5 m, and they cannot construct the reef crest to reach sea surface in the turbulent water conditions of breakwater.

Species-level investigation of Holocene coral reef facies in the northwest Pacific reefs revealed that the robust framework of the reef crest consists of a few species of coral: corymbose and tabular *Acropora* (*A. digitifera*, *A. hyacinthus*, and *A. robusta*/*A. abrotanoides*) and *Isopora palifera* (Hongo and Kayanne 2010, 2011; Hongo 2012). These corals distribute over a present reef edge to an upper reef slope, which is the growth frontier of the present reef formation. Therefore, the corymbose and tabular *Acropora* have a potential to maintain coral reef landforms to catch up with the rising sea level with a rate of 0.5 m/100 years at its maximum (Fig. 7.6). If sea level rise in this century tracks the medium scenario (RCP4.5 or RCP6.0), surface of the present reef crest would be able to catch up with the sea level to maintain its function of breakwater.

However, these key species are most vulnerable to thermal stresses and impact of typhoons (Hongo and Yamano 2013; Harii et al. 2014). For instance, super typhoons, which are projected to occur more frequently by the global warming, will dislodge *A. digitifera*, one of the key coral species (Hongo et al. 2012). Increasing both stresses will degrade the potential of the key species to form a rigid framework to catch up with the rising sea level.

7.4.2 Response of Atoll Islands

Atoll islands are areas of low, flat land, and the sustainability of habitable land in such environments is sensitive to even slight changes in sea level. In the tropical Pacific, a high sea stand of 1–2 m above present level was widely observed from 2000 to 4000 years before present, and many atoll islands were formed during the subsequent relative fall in sea level on the reef flat which caught up with this sea level (Schofield 1977; McLean and Woodroffe 1994; Woodroffe et al. 1999; Kayanne et al. 2011).

In Majuro Atoll, central Pacific, sea level reached high stand of 1.1 m above present mean sea level around 4000 years ago, and coral reef had been formed to catch up

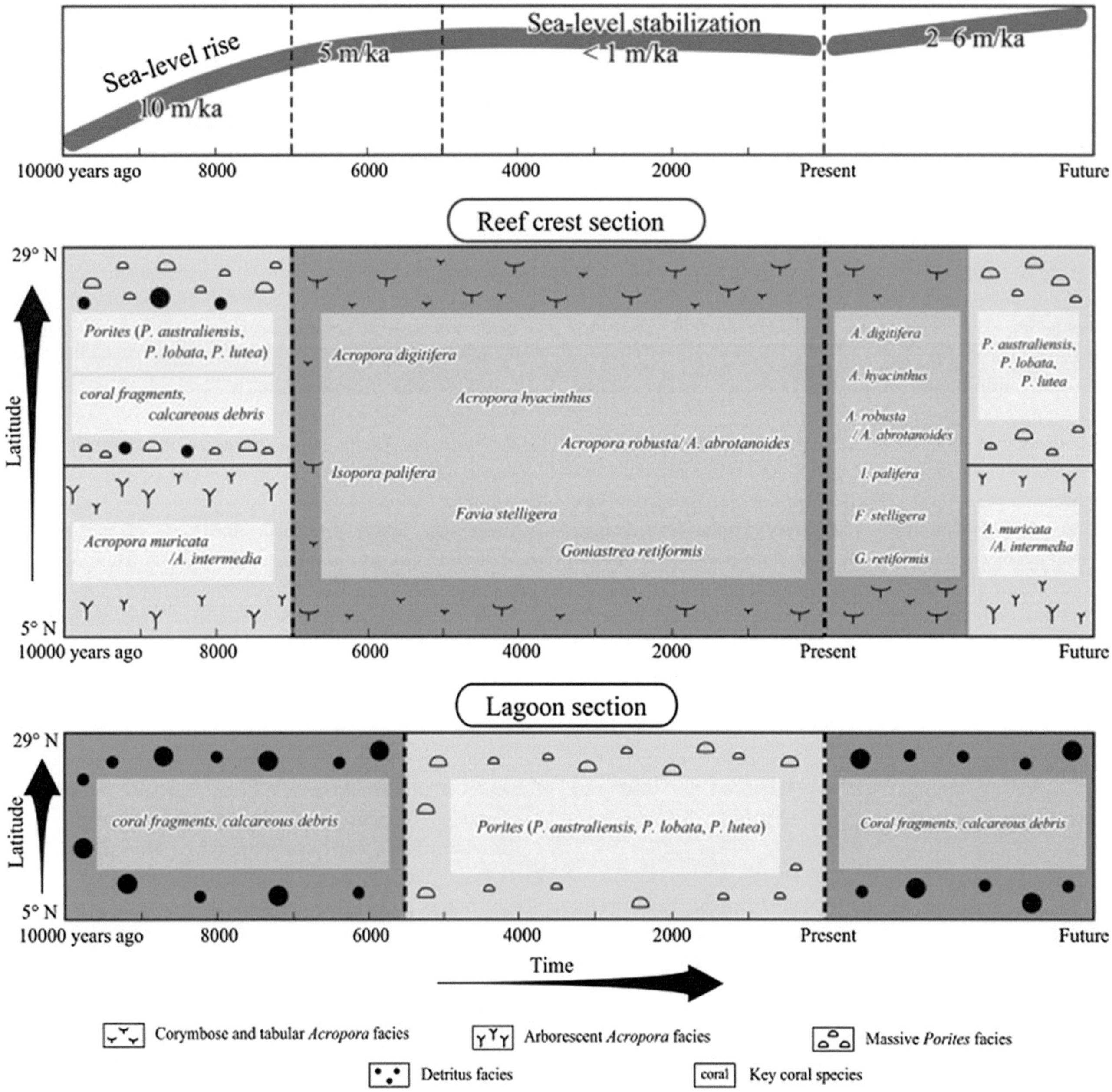

Fig. 7.6 Temporal and spatial patterns of sedimentary facies and key coral species in the Northwest Pacific (Hongo 2012). During the period of rapid sea level rise (10 m/ka) between 10,000 and 7000 years ago, arborescent *Acropora* (*A. muricata*/*A. intermedia*) and massive *Porites* were the dominant contributors to reef growth in tropical and subtropical regions, respectively. During the period of slower sea level rise (5 m/ka) between 7000 and 5000 years ago, the key species were completely replaced by *A. digitifera*, *A. hyacinthus*, *A. robusta*/*A. abrotanoides*, and *Isopora palifera* from corymbose and tabular *Acropora* facies. These coral species contributed to reef crest formation during the period of sea level stabilization to the present. The key coral species of the corymbose and tabular *Acropora* facies is expected to contribute to reef crest formation in response to future sea level rise at a rate of 2–6 m/ka (From Hongo (2012) with permission of Elsevier)

with this level (Fig. 7.7a). Then a fall of sea level occurred 2000 years ago, and an island consisting of coral gravel and foraminifera sand above high water level was formed within 100 years. The emergence of the island was followed by human settlement almost at once (Fig. 7.7b) (Kayanne et al. 2011). The narrow atoll islands have been continuously settled for the past 2000 years since the pioneering people migrated to colonize the bare island, which landscape has been modified by the settlers since colonization, with particular rapid change in recent years (Yamaguchi et al. 2009).

Fongafale Island, the capital of Tuvalu, inundated during spring high tide, which has been reported to be possibly

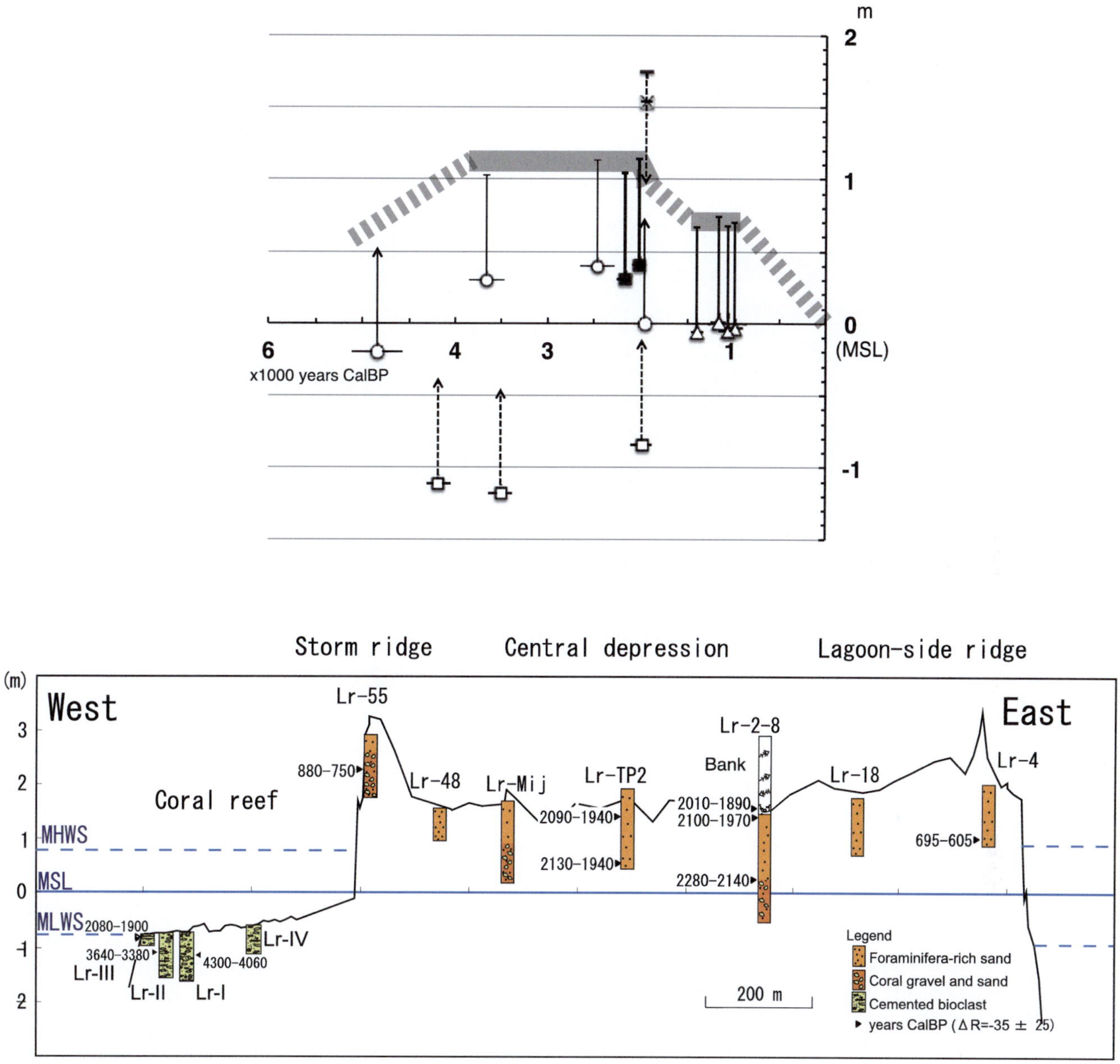

Fig. 7.7 Holocene sea level change and cross section of Laura Island, Majuro Atoll (From Kayanne et al. Rapid settlement of Majuro Atoll, central Pacific, following its emergence at 2000 years CalBP.)

related to sea level rise, since local people claim that they never had experienced such inundation before (Patel 2006). However, historical maps, aerial photos, and satellite images revealed that the inundation is resulted from expansion of residential area into low swampland in the course of population increase since the independence of Tuvalu in 1978 (Yamano et al. 2007).

Moreover, ecosystem degradation by human impact reduced the foraminifera sand production in the capital island of Majuro (Osawa et al. 2010) and Tuvalu (Fujita et al. 2013). The sand transportation along the coast was blocked by artificial constructions such as a causeway, piers, and dredges. Therefore, environmental problems in atoll islands at present are not as simple as submergence by sea level rise but more complex mainly induced by local human impacts in the course of population increase (Fig. 7.8). However, these local issues increased vulnerability of atoll islands against projected future sea level rise up to 1 m by

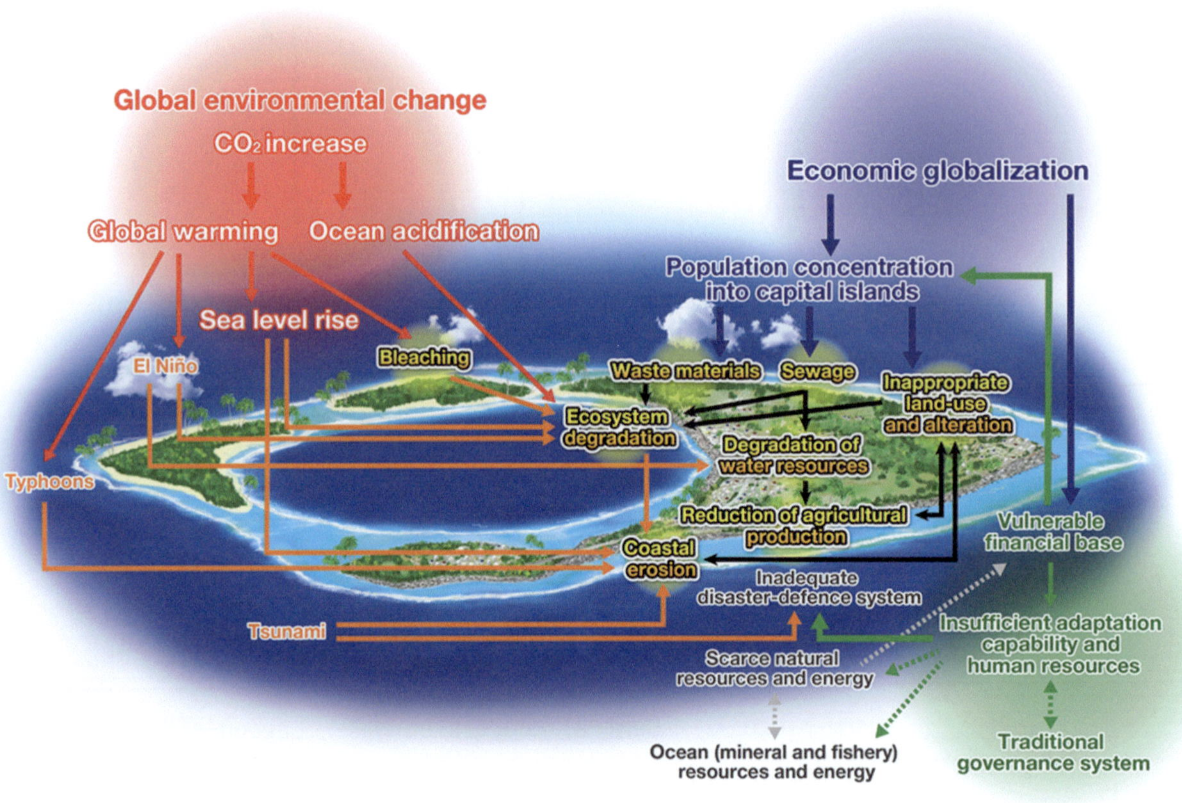

Fig. 7.8 Combined threats to atoll islands

the end of this century. Thus, to sustain atoll island against sea level rise, rehabilitation of natural island formation process is most important.

7.5 Management and Conservation of Coral Reefs in the "+2 °C World"

7.5.1 Future Scenario

Table 7.1 summarizes the future scenarios of the global warming relative to the preindustrial level, roughly approximating the RCP scenarios of AR5, IPCC. We have already emitted 370 GtC by fossil fuel burning and atmospheric CO_2 concentration reached 400 ppm. The averaged temperature has increased 0.8 °C, pH has decreased by 0.1 unit to 8.1, and sea level has risen by 0.2 m from preindustrial time. If we tackle with the global warming by reducing fossil fuel emission and by increasing CO_2 sink at once, the CO_2 concentration will be stabilized at 420 ppm, and there will be an increase of temperature at 1.5 °C (+1 °C world in Table 5.2) in the twenty-first century. The pH level will not reduce below 8.0, but sea level will show relatively large increase with a time lag to heat large volume of the ocean water and to melt glaciers. In the "+1 °C world," corals will be able to survive, but severe bleaching will occur every 5 years (Frieler et al. 2013). Ocean acidification will not affect most of the coral reefs, and they will be able to sustain coral reef landform with their reef crest keeping up with the rising sea level with 0.6 m/100 years. However, sustainability of coral reefs will be achieved only by healthy corals, and if humans would increase local pressure and degrade the reef health, they cannot sustain its landforms.

Under the condition of "+2 °C world," more probable scenario as we have shown no signature of reducing CO_2 emission by now, the future of coral reefs is more pessimistic. Severe bleaching would occur every year, and coral reef calcification would be reduced to become net dissolution below pH 8.0 unit. Reef crest cannot keep up with the rising sea level of 0.7 m, because the rate exceeds the potential vertical accumulation rate of coral reef top together with degradation of key coral species by warming and acidification. The coral population would shift from hard corals to macroalgae or soft corals, both of which have no function of reef landform formation.

Table 7.1 Approximate scenarios of the global warming by year 2100 relative to preindustrial time (year 1850)

		Cumulative CO_2 emission	Atmospheric CO_2 concentration	Average temperature	pH	Sea level (ultimate rise)	Corresponding RCP by IPCC
		GtC	ppm	°C		m	
Preindustrial		–	300	–	8.2	–	
Present		370	400	+0.8	8.1	+0.2	
In 2100	+1 °C world	650	420	+1.5	8.0	+0.6 (+1–2?)	RCP2.6
	+2 °C world	1000–1500	550–700	+2–+3	8.0–7.9	+0.7 (+2–4?)	RCP4.5, 6.0
	+4 °C world	2000	900	+4	7.8	+0.8 (+5–9?)	RCP8.5

In any cases, "+2 °C world" is the threshold to maintain coral reefs with their meaningful coverage of living corals. It might be possible for corals to adapt to the increasing SST and decreasing pH, yet it is susceptible that they adapt to the change within such a short time of 100 years. Under the worst scenario of "+4 °C world," no corals will be able to survive and present reef will be covered by macroalgae, which will be submerged by high rate of sea level rise in the centuries to come.

Each factor of the global warming would give a stress, and they combine to give a synergistic effect. Corals are more susceptible to bleaching by a combined effect of high SST and low pH (Anthony et al. 2008). If coral coverage and calcification rate would decrease by bleaching and ocean acidification, respectively, function of coral reefs to form a rigid structure to keep up with sea level rise would be lost.

7.5.2 Feedback Loops

Relation between the global warming scenario and coral reefs is schematically shown in Fig. 7.9. Solid line represents a positive coupling, in which an increase (decrease) in one component leads to an increase (decrease) in the linked component. When CO_2 increases, SST increases and sea level increases. Global warming scenario is linked by a series of positive couplings. In contrast, dashed line represents a negative coupling, in which an increase (decrease) in one component leads to a decrease (increase) in the linked component. As CO_2 increases, calcification decreases by the ocean acidification; and as SST increases, photosynthesis decreases by bleaching (Kayanne et al. 2005).

In some cases, link of the positive and negative couplings returned to the same component to form a positive or negative feedback loops. As CO_2 increases and as a result SST increases, photosynthetic production decreases leading to loss of fixation capacity of CO_2. This is the positive feedback loop as increase in CO_2 enhances its increase within the loop. On the other hand, as CO_2 increases, calcification decreases, which then buffers increase in CO_2, acting as a negative feedback loop. Increase in CO_2 enhances photosynthesis and fixes more CO_2 to counteracts its increase, which also acts as a negative feedback loop. All these loops are rather small to feedback to global environmental change, but would act to change coral reef carbonate chemistry (Fig. 7.5) (Anthony et al. 2011; Kleypas et al. 2011). Negative feedback loops act to stabilize reef water chemistry and buffer the increase of CO_2.

These feedback loops between the global environmental change and coral reefs are added to local human stresses such as eutrophication, effluent of silt, and over-fishing. These local stresses generally favor macroalgae dominance over corals and then degrade the coral reef landform formation processes. The system with feedback loops may sometimes trigger phase shift in which dramatic and irreversible state change occurs at a threshold point of stresses outside the system (Scheffer et al. 2001). Phase shift from coral-dominated to macroalgal-dominated reefs is regarded as one of the representative cases of ecosystem phase shift and has been modeled mainly between coral and macroalgal cover (Hoegh-Guldberg et al. 2007; Mumby et al. 2007). The increase or decrease of all the components can be reversed within the same positive or negative feedback loops. The decreasing trend of coral cover within a positive feedback loop can be reversed to increase its cover within the same positive feedback. The degradation of coral is not "negative" feedback as was misinterpreted in Mumby and Steneck (2008) and Mumby (2009).

7.5.3 Management and Conservation

Reduction of local stresses must firstly be achieved to conserve coral reefs against global environmental changes. If reef degradation was enhanced by a positive feedback loop, efforts to turn at least one component should be conducted to

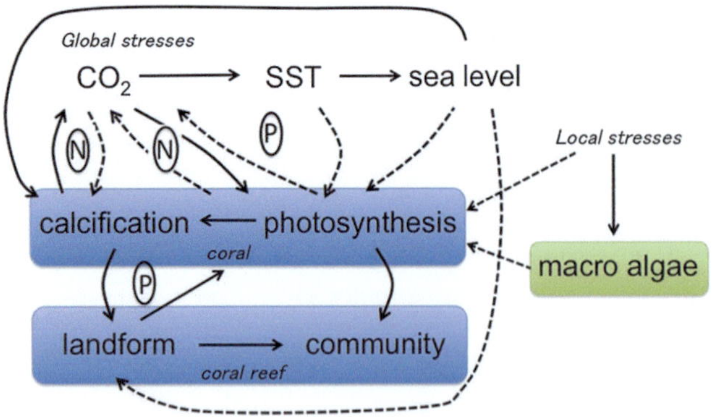

Fig. 7.9 Feedback loops between global stresses and coral and coral reefs. *Solid lines*: positive couplings; *dashed lines*: negative couplings. P and N denote positive and negative feedbacks, respectively

reverse it into reef rehabilitation: decrease macroalgal cover, reduction of local stresses or enhance coral metabolisms.

Magnitudes of effects of global warming and ocean acidification are varied geographically. Reef sites with higher rate of exchange with the ocean water or sheltered from warm surface water in relatively deep water must be prioritized to conserve and to remove local stresses.

Mass culture of corals must also be taken into consideration (Nakamura et al. 2011). In juvenile culture process, corals and symbiotic algae which are more tolerant to stresses should be challenged. In transplantation of juvenile corals, appropriate species must be transplanted at appropriate habitat. To maintain reef structure, the key coral species must be planted in the edge of reef crest. Removal of local stresses is prerequisite.

Adaptation of corals to global stresses (higher SST and lower pH) was not discussed in this chapter, though increasing number of studies have confirmed the evidence. However, it is susceptible that corals can keep up to adapt the enhanced rates of the changes in this century.

References

Anthony KRN, Kline DI, Diaz-Pulido G, Dove S, Hoegh-Guldberg O (2008) Ocean acidification causes bleaching and productivity loss in coral reef builders. Proc Natl Acad Sci U S A 105:17442–17446

Anthony KRN, Kleypas JA, Gattuso JP (2011) Coral reefs modify their seawater carbon chemistry – implications for impacts of ocean acidification. Glob Change Biol 17:3655–3666

Baker AC, Glynn PW, Riegl B (2008) Climate change and coral reef bleaching: an ecological assessment of long-term impacts, recovery trends and future outlook. Estuar Coast Shelf S 80:435–471

Burrows MT, Schoeman DS, Buckley LB, Moore P, Poloczanska ES, Brander KM, Brown C, Bruno JF, Duarte CM, Halpern BS, Holding J, Kappel CV, Kiessling W, O'Connor MI, Pandolfi JM, Parmesan C, Schwing FB, Sydeman WJ, Richardson AJ (2011) The pace of shifting climate in marine and terrestrial ecosystems. Science 334:652–655

Crook ED, Potts D, Rebolledo-Vieyra M, Hernandez L, Paytan A (2012) Calcifying coral abundance near low-pH springs: implications for future ocean acidification. Coral Reefs 31:239–245

Dadhich AP, Nadaoka K, Yamamoto T, Kayanne H (2012) Detecting coral bleaching using high-resolution satellite data analysis and 2-dimensional thermal model simulation in the Ishigaki fringing reef, Japan. Coral Reefs 31:425–439

Doney SC, Balch WM, Fabry VJ, Feely RA (2009) Ocean acidification: a critical emerging problem for the ocean sciences. Oceanography 22:16–25

Fabricius KE, Langdon C, Uthicke S, Humphrey C, Noonan S, De'ath G, Okazaki R, Muehllehner N, Glas MS, Lough JM (2011) Losers and winners in coral reefs acclimatized to elevated carbon dioxide concentrations. Nat Clim Chang 1:165–169

Frieler K, Meinshausen M, Golly A, Mengel M, Lebek K, Donner SD, Hoegh-Guldberg O (2013) Limiting global warming to 2°C is unlikely to save most coral reefs. Nat Clim Chang 3:165–170

Fujita M, Suzuki J, Sato D, Kuwahara Y, Yokoki H, Kayanne H (2013) Anthropogenic impacts on water quality of the lagoonal coast of Fongafale Islet, Funafuti Atoll, Tuvalu. Sustain Sci 8:381–390

Glynn PW (1993) Coral-reef bleaching – ecological perspectives. Coral Reefs 12:1–17

Hall-Spencer JM, Rodolfo-Metalpa R, Martin S, Ransome E, Fine M, Turner SM, Rowley SJ, Tedesco D, Buia MC (2008) Volcanic carbon dioxide vents show ecosystem effects of ocean acidification. Nature 454:96–99

Hamanaka N, Kan H, Yokoyama Y, Okamoto T, Nakashima Y, Kawana T (2012) Disturbances with hiatuses in high-latitude coral reef growth during the Holocene: correlation with millennial-scale global climate change. Glob Planet Chang 80–81:21–35

Harii S, Hongo C, Ishihara M, Ide Y, Kayanne H (2014) Impacts of multiple disturbances on coral communities at Ishigaki Island, Okinawa, Japan, during a 15 year survey. Mar Ecol Prog Ser 509:171

Hoegh-Guldberg O, Bruno JF (2010) The impact of climate change on the world's marine ecosystems. Science 328:1523–1528

Hoegh-Guldberg O, Mumby PJ, Hooten AJ, Steneck RS, Greenfield P, Gomez E, Harvell CD, Sale PF, Edwards AJ, Caldeira K, Knowlton N, Eakin CM, Iglesias-Prieto R, Muthiga N, Bradbury RH, Dubi A, Hatziolos ME (2007) Coral reefs under rapid climate change and ocean acidification. Science 318:1737–1742

Hongo C (2012) Holocene key coral species in the Northwest Pacific: indicators of reef formation and reef ecosystem responses to global climate change and anthropogenic stresses in the near future. Quat Sci Rev 35:82–99

Hongo C, Kayanne H (2010) Relationship between species diversity and reef growth in the Holocene at Ishigaki Island, Pacific Ocean. Sediment Geol 223:86–99

Hongo C, Kayanne H (2011) Key species of hermatypic coral for reef formation in the Northwest Pacific during Holocene sea-level change. Mar Geol 279:162–177

Hongo C, Yamano H (2013) Species-specific responses of corals to bleaching events on anthropogenically turbid reefs on Okinawa island, Japan, over a 15-year period (1995–2009). Plos One 8: e60952

Hongo C, Kawamata H, Goto K (2012) Catastrophic impact of typhoon waves on coral communities in the Ryukyu Islands under global warming. J Geophys Res Biogeol 117:G02029

Inoue S, Kayanne H, Yamamoto S, Kurihara H (2013) Spatial community shift from hard to soft corals in acidified water. Nat Clim Chang 3:683–687

Jokiel PL, Coles SL (1990) Response of Hawaiian and other Indo-Pacific reef corals to elevated-temperature. Coral Reefs 8:155–162

Kayanne H (1992) Deposition of calcium carbonate into Holocene reefs and its relation to sea-level rise and atmospheric CO2. In: Proceedings of the 7th international coral reef symposium 1:50–55

Kayanne H, Harii S, Ide Y, Akimoto F (2002) Recovery of coral populations after the 1998 bleaching on Shiraho Reef, in the Southern Ryukyus, NW Pacific. Mar Ecol Prog Ser 239:93–103

Kayanne H, Hata H, Kudo S, Yamano H, Watanabe A, Ikeda Y, Nozaki K, Kato K, Negishi A, Saito H (2005) Seasonal and bleaching-induced changes in coral reef metabolism and CO2 flux. Glob Biogeochem Cycles 19

Kayanne H, Yasukochi T, Yamaguchi T, Yamano H, Yoneda M (2011) Rapid settlement of Majuro Atoll, central Pacific, following its emergence at 2000 years CalBP. Geophysical Research Letters 38:L20405

Kleypas JA, Anthony KRN, Gattuso JP (2011) Coral reefs modify their seawater carbon chemistry – case study from a barrier reef (Moorea, French Polynesia). Glob Chang Biol 17:3667–3678

Liu G, Skirving WJ, Strong AE (2003) Remote sensing of sea surface temperatures during 2002 barrier reef coral bleaching. Eos Trans AGU 84:137–144

Liu G, Strong AE, Skirving WJ, Arzayus LF (2006) Overview of NOAA coral reef watch program's near-real-time satellite global coral bleaching monitoring activities. In: Proc 10th Int Coral Reef Symp 2, pp 1783–1793

McLean RF, Woodroffe CD (1994) Coral atolls. In: Carter B, Woodroffe CD (eds) Coastal evolution: late quaternary shoreline morphodynamics. Cambridge University Press, Cambridge, pp 267–302

Monastersky R (2013) Global carbon dioxide levels near worrisome milestone. Nature 497:13–14

Montaggioni LF (2005) History of Indo-Pacific coral reef systems since the last glaciation: development patterns and controlling factors. Earth-Sci Rev 71:1–75

Morse JW, Andersson AJ, Mackenzie FT (2006) Initial responses of carbonate-rich shelf sediments to rising atmospheric pCO(2) and "ocean acidification": role of high Mg-calcites. Geochim Cosmochim Ac 70:5814–5830

Moss RH, Edmonds JA, Hibbard KA, Manning MR, Rose SK, van Vuuren DP, Carter TR, Emori S, Kainuma M, Kram T, Meehl GA, Mitchell JFB, Nakicenovic N, Riahi K, Smith SJ, Stouffer RJ, Thomson AM, Weyant JP, Wilbanks TJ (2010) The next generation of scenarios for climate change research and assessment. Nature 463:747–756

Mumby PJ (2009) Phase shifts and the stability of macroalgal communities on Caribbean coral reefs. Coral Reefs 28:761–773

Mumby PJ, Steneck RS (2008) Coral reef management and conservation in light of rapidly evolving ecological paradigms. Trends Ecol Evol 23:555–563

Mumby PJ, Hastings A, Edwards HJ (2007) Thresholds and the resilience of Caribbean coral reefs. Nature 450:98–101

Nakamura R, Ando W, Yamamoto H, Kitano M, Sato A, Nakamura M, Kayanne H, Omori M (2011) Corals mass-cultured from eggs and transplanted as juveniles to their native, remote coral reef. Mar Ecol Prog Ser 436:161–168

Nakano Y (2004) Global environmental change and coral bleaching. In: Environment TJCRSaMot (ed) Coral reefs of Japan. Ministry of the Environment, Japan, pp 42–48

Nojima S, Okamoto M (2008) Enlargement of habitats of scleractinian corals to north and coral bleaching events. Nippon Suisan Gakkaishi 74:884–888

Osawa Y, Fujita K, Umezawa Y, Kayanne H, Ide Y, Nagaoka T, Miyajima T, Yamano H (2010) Human impacts on large benthic foraminifers near a densely populated area of Majuro Atoll, Marshall Islands. Mar Pollut Bull 60:1279–1287

Parmesan C, Yohe G (2003) A globally coherent fingerprint of climate change impacts across natural systems. Nature 421:37–42

Patel SS (2006) A sinking feeling. Nature 440:734–736

Paulay G, Benayahu Y (1999) Patterns and consequences of coral bleaching in Micronesia (Majuro and Guam) in 1992–1994. Micronesica 31:109–124

Plummer LN, Mackenzie FT (1974) Predicting mineral solubility from rate data – application to dissolution of magnesian calcites. Am J Sci 274:61–83

Poloczanska ES, Brown CJ, Sydeman WJ, Kiessling W, Schoeman DS, Moore PJ, Brander K, Bruno JF, Buckley LB, Burrows MT, Duarte CM, Halpern BS, Holding J, Kappel CV, O'Connor MI, Pandolfi JM, Parmesan C, Schwing F, Thompson SA, Richardson AJ (2013) Global imprint of climate change on marine life. Nat Clim Chang 3:919–925

Porter V, Leberer T, Gawel M, Gutierrez J, Burdick D, Torres V, Lujan E (2005) Status of the coral reef ecosystems of Guam University of Guam Marine Laboratory Technical Report, Guam 68p

Scheffer M, Carpenter S, Foley JA, Folke C, Walker B (2001) Catastrophic shifts in ecosystems. Nature 413:591–596

Schofield JC (1977) Effect of late Holocene sea-level fall on atoll development. N Z J Geol Geophys 20:531–536

Silverman J, Lazar B, Cao L, Caldeira K, Erez J (2009) Coral reefs may start dissolving when atmospheric CO2 doubles. Geophys Res Lett 36:GL036282

Sugihara K, Sonoda N, Imafuku T, Nagata S, Ibusuki T, Yamano H (2009) Latitudinal changes in hermatypic coral communities from West Kyushu to Oki Islands in Japan. J Jpn Coral Reef Soc 11:51–57

Takahashi A, Kurihara H (2013) Ocean acidification does not affect the physiology of the tropical coral Acropora digitifera during a 5-week experiment. Coral Reefs 32:305–314

van Woesik R, Sakai K, Ganase A, Loya Y (2011) Revisiting the winners and the losers a decade after coral bleaching. Mar Ecol Prog Ser 434:67–76

van Woesik R, Houk P, Isechal AL, Idechong JW, Victor S, Golbuu Y (2012) Climate-change refugia in the sheltered bays of Palau: analogs of future reefs. Ecol Evol 2:2474–2484

Veron JEN, Minchin PR (1992) Correlations between sea-surface temperature, circulation patterns and the distribution of hermatypic corals of Japan. Cont Shelf Res 12:835–857

Veron JEN, Hoegh-Guldberg O, Lenton TM, Lough JM, Obura DO, Pearce-Kelly P, Sheppard CRC, Spalding M, Stafford-Smith MG, Rogers AD (2009) The coral reef crisis: the critical importance of $<$ 350 ppm CO2. Mar Pollut Bull 58:1428–1436

Watanabe A, Yamamoto T, Nadaoka K, Maeda Y, Miyajima T, Tanaka Y, Blanco AC (2013) Spatiotemporal variations in CO2 flux in a fringing reef simulated using a novel carbonate system dynamics model. Coral Reefs 32:239–254

Wilkinson C (2002) Status of coral reefs of the world 2002. Australian Institute of Marine Science, Townsville

Wilkinson C, Linden O, Cesar H, Hodgson G, Rubens J, Strong AE (1999) Ecological and socioeconomic impacts of 1998 coral mortality in the Indian Ocean: an ENSO impact and a warning of future change? Ambio 28:188–196

Woodroffe CD, McLean RF, Smithers SG, Lawson EM (1999) Atoll reef-island formation and response to sea-level change: West Island, Cocos (Keeling) Islands. Mar Geol 160:85–104

Yamaguchi T, Kayanne H, Yamano H (2009) Archaeological investigation of the landscape history of an oceanic atoll: Majuro, Marshall islands. Pac Sci 63:537–565

Yamamoto S, Kayanne H, Terai M, Watanabe A, Kato K, Negishi A, Nozaki K (2012) Threshold of carbonate saturation state determined by CO2 control experiment. Biogeosciences 9:1441–1450

Yamamoto S, Kayanne H, Tokoro T, Kuwae T, Watanabe A (2015) Total alkalinity flux in coral reefs estimated from eddy covariance and sediment pore-water profiles. Limnol Oceanogr 60:229–241

Yamano H, Kayanne H, Yamaguchi T, Kuwahara Y, Yokoki H, Shimazaki H, Chikamori M (2007) Atoll island vulnerability to flooding and inundation revealed by historical reconstruction: Fongafale Islet, Funafuti Atoll, Tuvalu. Glob Planet Chang 57:407–416

Yamano H, Sugihara K, Nomura K (2011) Rapid poleward range expansion of tropical reef corals in response to rising sea surface temperatures. Geophys Res Lett 38:GL046474

Yara Y, Oshima K, Fujii M, Yamano H, Yamanaka Y, Okada N (2011) Projection and uncertainty of the poleward range expansion of coral habitats in response to sea surface temperature warming: a multiple climate model study. Galaxea J Coral Reef Stud 13:11–20

Yara Y, Vogt M, Fujii M, Yamano H, Hauri C, Steinacher M, Gruber N, Yamanaka Y (2012) Ocean acidification limits temperature-induced poleward expansion of coral habitats around Japan. Biogeosciences 9:4955–4968

Yoneyama S, Seno K, Yamamoto T (2008) Coral bleaching in Hahajima Island, Ogasawara Island chain in 2003. Tokyo Metrop Res Fish Sci 2:81–93

Atmospheric Deposition of Reactive Nitrogen as a Regional-Scale Eutrophication Stress on the Coral Reef Ecosystem

8

Toshihiro Miyajima, Naoko Morimoto, Takashi Nakamura, Takahiro Yamamoto, Atsushi Watanabe, and Kazuo Nadaoka

Abstract

Long-range transport and deposition of atmospheric pollutants from mid-latitude industrial regions to low-latitude seas have the potential to degrade coral reefs. We investigated the atmospheric wet deposition of dissolved inorganic nitrogen (DIN) and phosphorus (DIP) to coral reef sites around the Ishigaki and Iriomote Islands in the subtropical western North Pacific. The deposition rate of DIN was higher in autumn and winter than in summer. The annual N deposition was 3–8 times higher than rates previously observed at subtropical North Atlantic reef sites and was almost as large as the cyanobacterial N_2 fixation rate previously estimated for Ishigaki reefs. A backward trajectory analysis of an air mass suggested that the dominant remote emission source for atmospheric nitrate in winter was coastal industrial areas in continental China. A comparison with previous reports suggested that the influence of transboundary pollution on the N budget at the study site had significantly increased during the first decade of the twenty-first century.

Keywords

Atmospheric deposition • Coral reefs • Reactive nitrogen • Seasonal variation

Atmospheric deposition of biophilic elements has been recognized as one of the major factors responsible for coastal eutrophication worldwide. In particular, the atmospheric deposition of reactive nitrogen (N) species (NO_3^-, NH_4^+, and organic N) has vastly increased in recent decades, with the increasing application of N fertilizers derived from industrial N_2 fixation and worsening air pollution due to fossil fuel combustion in power plants and vehicles (Gruber and Galloway 2008). Terrestrial and coastal marine ecosystems are facing sustained and intensifying stress due to excess N loadings from atmospheric deposition, which potentially causes strong perturbations in both the availability and stoichiometry of inorganic nutrients for primary producers (Owens and Galloway 1992; Paerl et al. 2002).

Because the tropospheric residence time of aerosol containing reactive N species (N_r) is not long (~1 week, with the exception of peroxyacetylnitrate; Jacob 1999), the influence of atmospheric N_r deposition (ADN_r) cannot extend globally and is usually restricted to within the vicinity and downwind regions (within a few thousand kilometers) of major source areas such as North America, Europe, and East Asia (Galloway et al. 2004; Dentener et al. 2006; Doney et al. 2007; Kim et al. 2011). Consequently, it has been presumed that the ecological influence of

The original version of this chapter was revised. An erratum to this chapter can be found at DOI 10.1007/978-4-431-54364-0_9.

T. Miyajima (✉) • N. Morimoto
Atmosphere and Ocean Research Institute, The University of Tokyo, Kashiwanoha 5-1-5, Kashiwa, Chiba 277-8564, Japan
e-mail: miyajima@aori.u-tokyo.ac.jp

T. Nakamura • A. Watanabe • K. Nadaoka
School of Environment and Society, Tokyo Institute of Technology, Tokyo, Japan

T. Yamamoto
School of Environment and Society, Tokyo Institute of Technology, Tokyo, Japan

Environment and Life Science Center, Kuwait Institute for Scientific Research, Kuwait, Kuwait

H. Kayanne (ed.), *Coral Reef Science*, Coral Reefs of the World 5, DOI 10.1007/978-4-431-54364-0_8

ADN_r in the ocean would be significant in areas directly downwind of most populated and industrialized areas, e.g. the US Atlantic, the Baltic Sea, and the western Mediterranean (Fanning 1989; Paerl 1997). However, the application of industrially fixed N and fossil fuel combustion are now increasing dramatically in economically less-developed countries within the tropics and subtropics, which may generate significant ADN_r in low-latitude coastal marine ecosystems. Furthermore, because polluted air masses can be transported for long distances over the sea depending on the regional wind regime (Wenig et al. 2003), the northerly monsoon may effectively convey atmospheric N_r from mid-latitude pollution sources to low-latitude oceans. The potential ecological impact of an external N_r loading, including ADN_r, is particularly large in ecosystems that were originally N-limited, and N is clearly one of the most limiting nutrients in many coral reefs as indicated by the strikingly high cyanobacterial N_2 fixing activity over reefs (D'Elia and Wiebe 1990). Increasing ADN_r potentially causes significant and irreversible changes in the ecological and biogeochemical performance of coral reef ecosystems.

Atmospheric N_r deposition may occur as wet and dry deposition, with the dry deposition further divided into the fallout of particulate matter and gaseous N (NH_3) deposition. Gaseous NH_3 deposition is intensive only in the vicinity of local NH_3 sources and rapidly diminishes with increasing distance (Aneja et al. 2001). In coastal areas, wet deposition and particulate dry deposition are the major components of ADN_r. The relative contribution of wet and dry N_r deposition can vary by site, but wet deposition is usually the dominant fraction of ADN_r, except in arid climates or under drought conditions (Meyers et al. 2001; Poor et al. 2001; Gao 2002). Consequently, the magnitude of ADN_r is constrained by the amount of rainfall as well as the positional relation to the source sites of atmospheric N_r. The areal extent of potential N_r sources to a particular recipient site of ADN_r, often referred to as an "airshed" (Paerl 2002), obviously depends on the wind regime and may vary between seasons where the monsoon has a strong influence. Such seasonal variation may enhance or reduce the ecological impact of ADN_r depending on the phenology of resident organisms.

Because ADN_r is supplied widely and exclusively to the sea surface, it will stimulate the primary production of phytoplankton rather than benthic primary producers, including reef corals (Paerl et al. 2002). The influence of ADN_r will then propagate to the grazers of phytoplankton and may improve the survival of planktonic larvae of benthic invertebrates if their occurrence synchronizes with a major ADN_r event. Such a cascading effect would be particularly important when considering the influence of ADN_r at the community level. A higher survival rate of planktonic coral larvae would support the health and resilience of coral populations. In contrast, enhanced larval survival of coral grazers such as *Acanthaster planci* may result in deterioration of the community. Therefore, to understand and predict the ecological influence of ADN_r on coral reefs, we must not only evaluate the bulk amount of ADN_r but also identify the dominant factors that govern both the spatial and temporal variability of ADN_r.

To achieve this, we investigated a case study of the wet deposition of N_r to coral reefs around the Yaeyama Islands in the subtropical western North Pacific, where the strong influence of the East-Asian monsoon and frequent tropical storms generates a drastic temporal variation in the airshed (see Fig. 8.4 for the location). We performed frequent field trips from March 2009 to September 2012, most intensively in winter (January) and summer (August/September), while short-term investigations were also conducted in other periods. We collected rain water samples for every rainfall event during each survey period and analyzed the samples for pH, electric conductivity, nutrients (NO_3^-, NO_2^-, NH_4^+, and PO_4^{3-}) and major ion concentrations, and δ^2H and $\delta^{18}O$ of water, depending on the amount of sample collected. We also attempted to localize the source site of N_r for each rainfall event by a backward trajectory analysis with the aid of the National Oceanic and Atmospheric Administration (NOAA) Hybrid Single Particle Lagrangian Integrated Trajectory (HYSPLIT) Model (www.arl.noaa.gov).

The subtropical climate of the Yaeyama Islands is illustrated by the clear seasonal change of temperature (16.6–29.4 °C: monthly mean air temperature), while rainfall occurs year round (Fig. 8.1a). Occasionally, intense rainfall is received due to tropical storms, especially from July to October. The relationship between the oxygen ($\delta^{18}O$) and hydrogen (δ^2H) isotope ratios of rain water differed in winter (January) and the warmer months (April–September, Fig. 8.2); specifically, the deuterium excess (*d*, defined as $\delta^2H - 8\ \delta^{18}O$) was much higher in January (20–30) than between April and September (4–14). This demonstrates that the monsoon is the major source and transport process of water vapor for rainfall in the study area (Araguás-Araguás et al. 1998). In winter, the northerly monsoon generated by the Siberian boreal air mass promotes rapid evaporation from the warm seawater of the East China Sea, resulting in water vapor labelled with high *d*, which is further transported to the Yaeyama Islands, where it causes rainfall. In the warmer season, water vapor formed under near-equilibrium conditions (consequently labelled with lower *d*) over the Philippine Sea or the South China Sea is predominantly transported by the southerly monsoon generated by the North Pacific tropical anticyclone and causes rainfall in the study area. Heavy rainfall brought about by typhoons often had extremely negative $\delta^{18}O$ and δ^2H (Fig. 8.2), but the *d* value was similar to those observed in the warmer months.

The volume-weighted monthly average concentrations of NO_3^-, NH_4^+, and PO_4^{3-} displayed large temporal variation

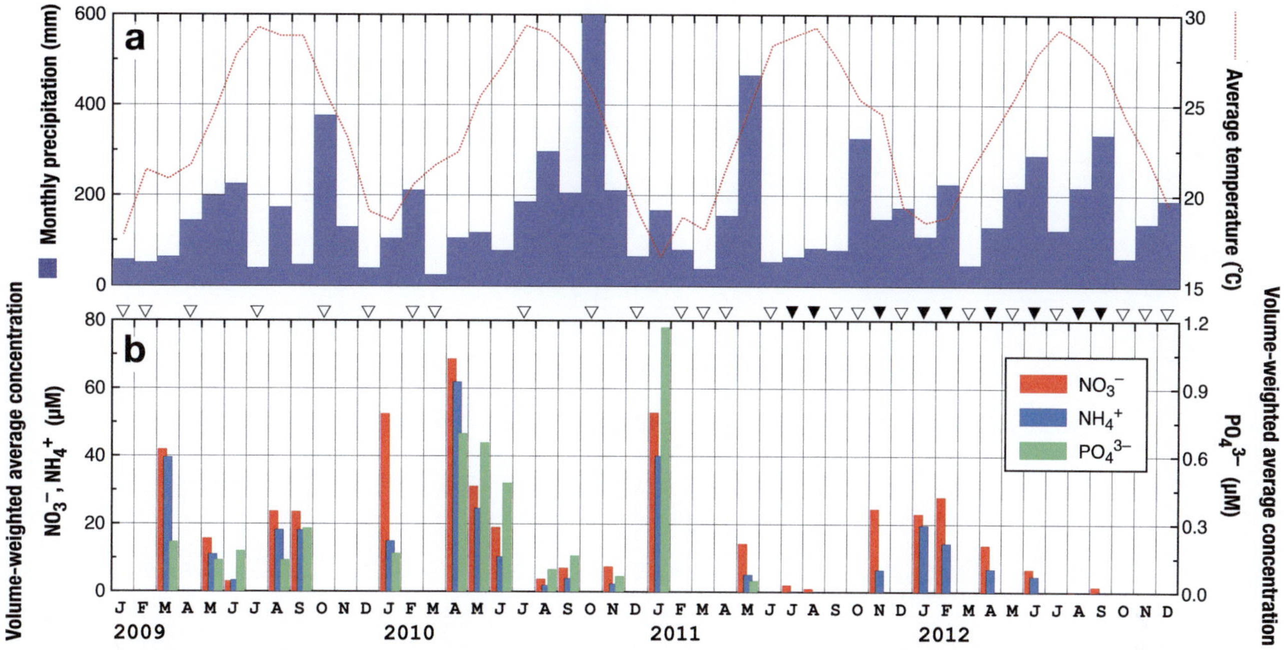

Fig. 8.1 (**a**) Monthly precipitation and average air temperature at Ishigaki Island (the main island of Yaeyama; published by the Japan Meteorological Agency) and (**b**) volume-weighted monthly average concentrations of NO_3^-, NH_4^+, and PO_4^{3-} in rain water collected at the Yaeyama Islands from January 2009 to September 2012. The open triangles above (**b**) indicate the months in which no rain water samples were collected; closed triangles indicate the months in which NO_3^- and NH_4^+, but not PO_4^{3-}, were determined

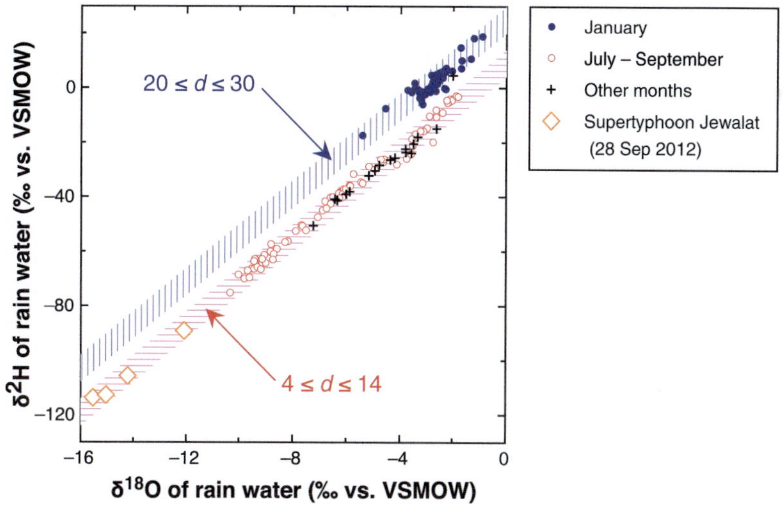

Fig. 8.2 Oxygen ($\delta^{18}O$) and hydrogen ($\delta^{2}H$) isotope ratios of rain water samples collected between November 2009 and September 2012. The isotope ratios are represented by the per-mill deviation from the international standard Vienna Standard Mean Ocean Water (VSMOW)

(Fig. 8.1b). While annual variation was significant, concentrations were always higher in winter (January to March) than summer (June to September) when comparing successive summer and winter seasons. The pH value of rain water was also clearly different between winter (mostly 4.0–5.0) and summer (>5.0). The seasonal changes in monthly average concentration were similar between NO_3^-, NH_4^+, and PO_4^{3-} (Fig. 8.1b). However, when the concentrations were compared for individual rainfall events, the correlation between these species was not necessarily strong (Fig. 8.3). The ratio of NH_4^+ to NO_3^- was usually less than 1 (0.86 on average), although significant deviation from this average was often observed (Fig. 8.3a). The ratio of PO_4^{3-} to NO_3^- was even more variable, being <0.02 in most cases, but was generally lower in winter than summer (Fig. 8.3b). There was a strong correlation between NH_4^+ and non-sea-salt-SO_4^{2-} ($r = 0.7405$, $p < 0.0001$; Fig. 8.3c), which indicates that NH_4^+ was transported

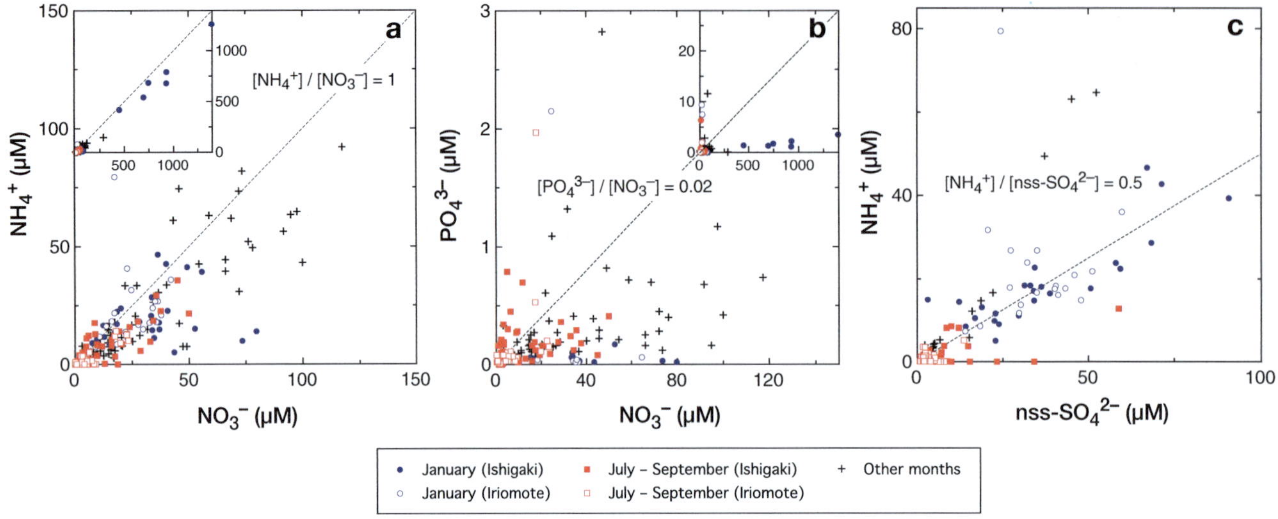

Fig. 8.3 Concentrations of (**a**) NH_4^+ vs. NO_3^-, (**b**) PO_4^{3-} vs. NO_3^-, and (**c**) NH_4^+ vs. non-sea-salt SO_4^{2-} (nss-SO_4^{2-}) in rain water samples collected between March 2009 and January 2012

through the atmosphere principally in the form of sulfate and hydrogensulfate aerosol (Nakamura et al. 2005). The above observations suggest that the flux of these important nutrients transported through the atmosphere was largely determined by the seasonal cycle of the large-scale wind regime and precipitation, although there was a significant short-term fluctuation in concentrations, presumably due to the variable efficiency of nucleation and entrapping aerosol into individual rainfall events.

During the intensive campaigns conducted in January and August/September, we collected rain water samples at five sites around Ishigaki Island and another five sites around Iriomote Island (ca. 20 km west of Ishigaki Island). The results confirmed that significant spatial variation in nutrient concentrations was present even in a single rainfall event. The amount of precipitation was often higher at Iriomote than Ishigaki Island, presumably reflecting the more mountainous landform of the former. Nutrient and major ion concentrations in rain water were often variable among the sites at each of the islands, and a general trend was observed where the nutrient concentrations in a single rainfall event were higher on the windward side of the island and/or where the amount of precipitation was smaller. This trend was likely related to the fact that nutrient species tend to be concentrated in the earlier rainfall during an individual rain event (Seymour and Stout 1983; Radojevic and Lim 1995). This observation suggests that the spatial pattern of ADN_r can be heterogeneous on a scale of kilometers, which is potentially of importance when considering the ecological consequences of ADN_r. This also indicates the difficulty in quantitatively evaluating the ADN_r to coastal ecosystems by single-point monitoring.

A backward trajectory simulation using the HYSPLIT model for individual rainfall events revealed that rainfall observed in summer at the Yaeyama Islands was usually caused by water vapor transported from the Philippine Sea (e.g., the rainfall events on 23 August 2010 and 11 July 2011 in Fig. 8.4) and occasionally from the South China Sea. In such cases, when the air mass had resided over the Philippine Sea or the South China Sea for at least three days before it caused rainfall on the Yaeyama Islands, the concentrations of NO_3^- and NH_4^+ in the rain water were low (<10 μmol L^{-1}). In contrast, rainfall in winter was usually caused by water vapor transported across the East China Sea. In particular, the concentrations of NO_3^- and NH_4^+ in rain water were extremely elevated when the air mass had passed over the industrial area of northeast China (including Beijing and Tianjin) within three days before it brought about precipitation at the study sites. The case of 29 January 2011 in Fig. 8.4 is a good example, with 420–1400 μmol L^{-1} of NO_3^- and NH_4^+ detected in rainwater samples collected at four sites on Ishigaki Island. When the air mass originated from the west (e.g., Taiwan and the Yangtze River Delta), the nutrient concentrations in rain water (usually 20–100 μmol L^{-1} NO_3^- and NH_4^+) were not as pronounced, but were still significantly higher than in normal rain water in summer. However, even in summer, rainfall is occasionally received from clouds that pass across the East China Sea, and in such cases, the nutrient concentrations in rain water can be elevated to the levels experienced in winter. In summer of 2009, the center of the Pacific tropical anticyclone was positioned further east than in a normal year. Tropical storms were frequently formed and proceeded north across the Philippine Sea. Consequently, wind often blew from the north to the

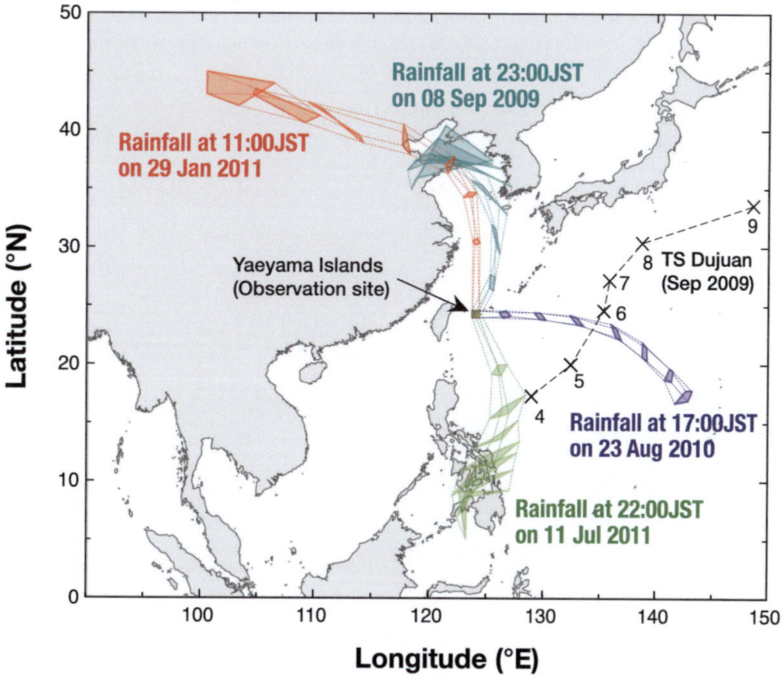

Fig. 8.4 Simulated trajectories of air masses that caused rainfall at the Yaeyama Islands for four typical cases (see text). An air mass that was positioned over a rectangle covering the Yaeyama Islands ($6' \times 6'$) at the time of the rainfall event was tracked backward every 12 hours (h) over an 84-h period before the rainfall, which is represented by tetragons. The trajectory of each corner of the rectangle was estimated from a backward trajectory simulation using the National Oceanic and Atmospheric Administration (NOAA) Hybrid Single Particle Lagrangian Integrated Trajectory (HYSPLIT) Model (www.arl.noaa.gov), with default settings adopted and the destination height set at 500 m above mean sea level. The track of Tropical Storm Dujuan, which moved across the Philippine Sea in September 2009 is also shown

Yaeyama Islands, causing rainfall that was significantly enriched with nutrients (see the case of 8 September 2009 and the track of TS Dujuan in Fig. 8.4 as an example, when rainwater NO_3^- and NH_4^+ were 44–92 μmol L^{-1} on Ishigaki Island). As a result, the volume-weighted mean nutrient concentrations of rain water in August and September of 2009 were considerably higher than those in the same months of 2010–2012 (Fig. 8.1b).

These observations confirmed that the wet deposition of nutrients to a particular recipient area is largely determined by the regional (*not* local) wind regime a few days before the rainfall event, and thus is more or less predictable from the backward trajectory simulation of an air mass once the major pollution sources are located. The annual variation in ADN_r may also be predicted qualitatively from the strength and duration of the monsoon.

We calculated the volume-weighted average concentrations of NO_3^- and NH_4^+ in rain water for each month using the data between March 2009 and January 2012, and multiplied them with the mean monthly precipitation at Ishigaki Island for 2003–2010 published by the Japan Meteorological Agency, to estimate the mean annual wet deposition of these species. The annual wet deposition of NO_3^- and NH_4^+ was 35.7 and 21.9 mmol m^{-2} $years^{-1}$, respectively. These values were slightly higher for NO_3^- and lower for NH_4^+ than previous estimates based on the precipitation data at Ishigaki Island for 1997–1998 (Agata et al. 2006), and about twice as high as the NO_3^- and slightly higher than the NH_4^+ estimates made at Okinawa Island in 2000 (Tomoyose et al. 2003; Table 8.1). Interestingly, Agata et al. (2006) reported that the concentration of NO_3^- in rain water was relatively low in winter (December to February), which contrasted with our results. They also found that peaks of NH_4^+ deposition appeared in June, October, and January, which they ascribed to local sources (i.e., ammonium fertilizer application). Tomoyose et al. (2003) showed a winter peak of wet deposition of NO_3^- at Okinawa Island (ca. 380 km northeast of Ishigaki Island) since 1996; the deposition gradually increased from 1996 to 2001, presumably due to increasing transboundary pollution from continental China (*cf.* Ohara et al. 2007). They also reported that the peak of NH_4^+ deposition occurred in summer and ascribed it to local sources. Thus, the transboundary NO_3^- pollution conveyed by the winter monsoon seems to have been evident at Okinawa Island (26.4°N), but not at Ishigaki Island (24.4°N) at the end of the last century. However, the transport of atmospheric NO_3^- from continental source areas in winter seems to have been significantly increased during the last decade due to the increasing industrial activity in northern China, as suggested by Tomoyose

Table 8.1 Annual wet deposition rates of reactive nitrogen (N) measured on islands in the subtropical North Atlantic and western North Pacific. The deposition rates of atmospheric reactive N in the potential source area (northeast China) for the subtropical western North Pacific in winter are also listed for comparison. The numbers in parentheses indicate the range of monthly variation

		Annual wet deposition (mmol N m^{-2} year^{-1})				
	Period	NO_3^-	NH_4^+	DIN	TN	Reference
North Atlantic						
Sargasso Sea	1982–1984	2.2 ~ 10.6	3.7 ~ 7.7	11.0	5.8 ~ 29.2	Knap et al. (1986)
Bahamas	2000	4.3	3.7	8.0	–	Barile and Lapointe (2005)
Bermuda	2008	5.6	5.2 ~ 10[a]	–	10 ~ 19	Knapp et al. (2010)
Subtropical Northwestern Pacific						
Ishigaki Island	1997–1998	34 (6 ~ 98)	32 (5 ~ 75)	66	–	Agata et al. (2006)
Okinawa Island	2000	13.2 (2.9 ~ 26.5)	18.5 (0 ~ 46.3)	31.7	–	Tomoyose et al. (2003)
Ishigaki and Iriomote Islands	2009–2011	35.7 (2.7 ~ 55.3)	21.9 (0.1 ~ 52.2)	57.5	–	This study
Islands around Korean Peninsula	2002–2008	–	–	53 ~ 110[b]	–	Kim et al. (2011)
Potential source area (NE China)	2007–2010	35 ~ 62	79 ~ 146	116 ~ 202	–	Pan et al. (2012)

[a]Including organic nitrogen
[b]Including dry deposition

et al. (2003), which could be the reason why a winter peak of NO_3^- was observed in this study, but not in Agata et al. (2006). On the other hand, the inconsistency of the seasonal variation in the wet deposition of NH_4^+ between these previous studies and our data was primarily due to differences in the collection sites of rain water. Both Tomoyose et al. (2003) and Agata et al. (2006) monitored and collected wet deposition at a single point where the influence of local agricultural and/or urban sources of atmospheric N_r was apparently high. In contrast, we established several collection sites of rain water, some of which were facing the sea and the others in mountainous areas. The contribution of local sources to these sites should have been relatively low.

For comparison, some previous studies of ADN_r at coral reef sites in the subtropical North Atlantic are listed in Table 8.1. The continental US is considered to be the principal source of ADN_r to this region (Barile and Lapointe 2005). The annual ADN_r is clearly much larger in the western North Pacific than in North Atlantic. This difference may be partly due to the difference in the local agricultural activity and partly to the difference in the magnitude of industrial and agricultural N_r emissions between the USA and China. However, the much stronger seasonal variations in ADN_r in the western North Pacific suggest that the East Asian monsoon plays a key role in determining the magnitude of ADN_r in the downwind coastal areas. This comparison illustrates that there is a large difference in the potential impact of ADN_r on coral reefs induced by regional-scale meteorological structure.

What ecological impact could the ADN_r exert on the coral reef ecosystems? With regard to the N budget, the answer may be provided by comparing the magnitude of ADN_r with other potential N_r sources. Umezawa et al. (2002) estimated N_r inputs by cyanobacterial N_2 fixation and terrestrial loading (groundwater discharge) in two coral reefs around Ishigaki Island with contrasting groundwater fluxes. The cyanobacterial N_2 fixation was projected to be between 40 and 90 mmol N m^{-2} year^{-1} in both reefs. The terrestrial N_r loading was estimated to be 30–170 and 390–550 mmol N m^{-2} year^{-1} in the groundwater-poor (Kabira) and rich (Shiraho) reefs, respectively. A comparison of these figures with those in Table 8.1 implies that the ADN_r could make a significant contribution to the N_r budget of coral reef ecosystems, being comparable to the cyanobacterial N_2 fixation, at least in coral reefs where the terrestrial N_r loading is relatively low. It is important to note that the estimation of ADN_r in Table 8.1 considered only the wet deposition settling directly on the water surface of the reef area. The terrestrial N_r loading estimated by Umezawa et al. (2002) included part of the ADN_r to the watershed in addition to N_r of local origin. The ADN_r to the outer ocean can also contribute to the reef N_r budget if it is transported by advection to the reef area. Therefore, the actual contribution of ADN_r to the reef N_r budget may be much greater than this comparison suggests.

Furthermore, the ADN_r can exert an ecological influence in a different way than the terrestrial loading and N_2 fixation. Because the terrestrial N_r is supplied by river discharge through the estuary and/or groundwater emission from a sandy beach, it is initially utilized by the estuarine and/or sand-flat communities. N_2 fixation is mainly promoted by benthic cyanobacteria in coral reefs; thus, the fixed N initially benefits the benthic primary producers and their grazers. In contrast, the ADN_r primarily supplies N_r to the water-column primary producers such as phytoplankton in the reef area and the outer ocean. Therefore, it is essential to understand the dynamics of suspended particulate organic matter including phytoplankton to determine how the ADN_r is connected to the nutrition of benthic consumers in coral reefs. The ADN_r may also provide N_r to reef coral species that are specialized in acquiring dissolved inorganic nutrients at extremely low concentrations from the water column, such as acroporid corals residing on the shallow reef flat. Thus, there are a number of factors that require clarification before we can correctly understand and predict the ecological consequences of ADN_r in coral reef ecosystems. This will require further intensive studies, especially experimental and modeling approaches using advanced tools, such as stable isotopes.

References

Agata S, Kumada M, Satake H (2006) Characteristics of hydrogen and oxygen isotopic compositions and chemistry of precipitation on Ishigaki Island in Okinawa, Japan. Chikyukagaku (Geochemistry) 40:111–123 (in Japanese with English summary)

Aneja VP, Roelle PA, Murray GC, Southerland J, Erisman JW, Fowler D, Asman WAH, Patni N (2001) Atmospheric nitrogen compounds II: emissions, transport, transformation, deposition and assessment. Atmos Environ 35:1903–1911

Araguás-Araguás L, Froehlich K, Rozanski K (1998) Stable isotope composition of precipitation over Southeast Asia. J Geophys Res 103:721–728

Barile P, Lapointe B (2005) Atmospheric nitrogen deposition from a remote source enriches macroalgae in coral reef ecosystems near Green Turtle Cay, Abacos, Bahamas. Mar Pollut Bull 50:1262–1272

D'Elia CF, Wiebe WJ (1990) Biogeochemical nutrient cycles in coral-reef ecosystems. In: Dubinsky Z (ed) Coral reefs. Elsevier Science Publishers B.V, Amsterdam, pp 49–74

Dentener F., Drevet J., Lamarque J.F., Bey I., Eickhout B., Fiore A.M., Hauglustaine D., Horowitz L.W., Krol M., Kulshrestha U.C., Lawrence M., Galy-Lacaux C., Rast S., Shindell D., Stevenson D., Van Noije T., Atherton C., Bell N., Bergman D., Butler T., Cofala J., Collins B., Doherty R., Ellingsen K., Galloway J., Gauss M., Montanaro V., Muller J.F., Pitari G., Rodriguez J., Sanderson M., Solmon F., Strahan S., Schultz M., Sudo K., Szopa S. and Wild O. (2006) Nitrogen and sulfur deposition on regional and global scales: a multimodel evaluation. Global Biogeochem Cycles 20, doi:10.1029/2005GB002672

Doney SC, Mahowald N, Lima I, Feely RA, Mackenzie FT, Lamarque J-F, Rasch PJ (2007) Impact of anthropogenic atmospheric nitrogen and sulfur deposition on ocean acidification and the inorganic carbon system. Proc Natl Acad Sci U S A 104:14580–14585

Fanning K (1989) Influence of atmospheric pollution on nutrient limitation in the ocean. Nature 339:460–463

Galloway J, Dentener F, Capone D, Boyer E, Howarth R, Seitzinger S, Asner G, Cleveland C, Green P, Holland E, Karl D, Michaels A, Porter J, Townsend A, Vorosmarty C (2004) Nitrogen cycles: past, present, and future. Biogeochemistry 70:153–226

Gao Y (2002) Atmospheric nitrogen deposition to Barnegat Bay. Atmos Environ 36:5783–5794

Gruber N, Galloway J (2008) An Earth-system perspective of the global nitrogen cycle. Nature 451:293–296

Jacob DJ (1999) Introduction to Atmospheric Chemistry. Princeton University Press, Princeton

Kim TW, Lee K, Najjar RG, Jeong HD, Jeong HJ (2011) Increasing N abundance in the northwestern Pacific Ocean due to atmospheric nitrogen deposition. Science 334:505–509

Knap A, Jickells T, Pszenny A, Galloway J (1986) Significance of atmospheric-derived fixed nitrogen on productivity of the Sargasso Sea. Nature 320:158–160

Knapp A, Hastings M, Sigman D, Lipschultz F, Galloway J (2010) The flux and isotopic composition of reduced and total nitrogen in Bermuda rain. Mar Chem 120:83–89

Meyers T, Sickles J, Dennis R, Russell K, Galloway J, Church T (2001) Atmospheric nitrogen deposition to coastal estuaries and their watersheds. In: Nitrogen loading in coastal water bodies: an atmospheric perspective. American Geophysical Union, Washington, pp 53–76

Nakamura T, Matsumoto K, Uematsu M (2005) Chemical characteristics of aerosols transported from Asia to the East China Sea: an evaluation of anthropogenic combined nitrogen deposition in autumn. Atmos Environ 39:1749–1758

Ohara T, Akimoto H, Kurokawa J-I, Horii N, Yamaji K, Yan X, Hayasaka T (2007) An Asian emission inventory of anthropogenic emission sources for the period 1980–2020. Atmos Chem Phys 7:4419–4444

Owens NJP, Galloway JN (1992) Episodic atmospheric nitrogen deposition to oligotrophic oceans. Nature 357:397–399

Paerl HW (1997) Coastal eutrophication and harmful algal blooms: importance of atmospheric deposition and groundwater as "new" nitrogen and other nutrient sources. Limnol Oceanogr 42:1154–1165

Paerl HW (2002) Connecting atmospheric nitrogen deposition to coastal eutrophication. Environ Sci Technol 36:323A–326A

Paerl HW, Dennis RL, Whitall DR (2002) Atmospheric deposition of nitrogen: implications for nutrient over-enrichment of coastal waters. Estuaries 25:677–693

Pan Y, Wang Y, Tang G, Wu D (2012) Wet and dry deposition of atmospheric nitrogen at ten sites in Northern China. Atmos Chem Phys 12:6515–6535

Poor N, Pribble R, Greening H (2001) Direct wet and dry deposition of ammonia, nitric acid, ammonium and nitrate to the Tampa Bay Estuary, FL, USA. Atmos Environ 35:3947–3955

Radojevic M, Lim L (1995) Short-term variation in the concentration of selected ions within individual tropical rainstorms. Water Air Soil Pollut 85:2363–2368

Seymour MD, Stout T (1983) Observations on the chemical composition of rain using short sampling times during a single event. Atmos Environ 17:1483–1487

Tomoyose N, Yogi K, Nagamine K (2003) Wet deposition in Ozato, Okinawa. Annual Report of Okinawa Prefectural Institute of Health and Environment 37, 47–50 (in Japanese)

Umezawa Y, Miyajima T, Kayanne H, Koike I (2002) Significance of groundwater nitrogen discharge into coral reefs at Ishigaki Island, southwest of Japan. Coral Reefs 21:346–356

Wenig M, Spichtinger N, Stohl A, Held G, Beirle S, Wagner T, Jähne B, Platt U (2003) Intercontinental transport of nitrogen oxide pollution plumes. Atmos Chem Phys 3:387–393

Erratum to: Coral Reefs of the World

Hajime Kayanne

H. Kayanne (ed.), *Coral Reef Science*, Coral Reefs of the World 5, DOI 10.1007/978-4-431-54364-0

DOI 10.1007/978-4-431-54364-0
DOI 10.1007/978-4-431-54364-0_1
DOI 10.1007/978-4-431-54364-0_2
DOI 10.1007/978-4-431-54364-0_7
DOI 10.1007/978-4-431-54364-0_8

The following corrections were made after original publication

Page iv, cover caption: Bleached coral colony (Porites cyrindrica) changed to Bleached coral colony (*Porites cylindrica*)

Page 6, line 4 (left column) and page 24 line 6 (right column): Wewenkang changed to Wewengkang

Page 22, line 19 (right column): Platax 12: 1–9 changed to Platax 12: 19–27

Page 26, lines 9 and 10 (right column): bad line break adjusted to keep "10" and "^{4}" together to read as "10^{4}"

Page 32, 4th and 5th lines from the bottom (left column): bad line break adjusted to keep "ml" and "$^{-1}$" together to read as "ml^{-1}"

Page 35, line 2 (left column): bad line break adjusted to keep "h" and "$^{-1}$" together to read as "h^{-1}"

Page 84, 3rd line from the bottom (left column), '55-year survey' was changed to '15-year survey'

The updated original online version for this book can be found at
DOI 10.1007/978-4-431-54364-0
DOI 10.1007/978-4-431-54364-0_1
DOI 10.1007/978-4-431-54364-0_2
DOI 10.1007/978-4-431-54364-0_7
DOI 10.1007/978-4-431-54364-0_8

H. Kayanne
Department of Earth and Planetary Science, Graduate School of Science, The University of Tokyo, Tokyo, Japan
e-mail: kayanne@eps.s.u-tokyo.ac.jp

H. Kayanne (ed.), *Coral Reef Science*, Coral Reefs of the World 5, DOI 10.1007/978-4-431-54364-0_9

Page 84, Fig. 7.3 caption revised to read as 'Locality map for examining relation between bleaching events and SST anomalies. Isolines show number of coral genera'

Page 92, Fig. 7.9 caption revised to read as 'Feedback loops between global stresses and coral and coral reefs. *Solid lines*: positive couplings; *dashed lines*: negative couplings. P and N denote positive and negative feedbacks, respectively'

Page 96, lines 33 and 34 (right column), bad line break adjusted to keep "δ^2" and "H" together to read as "δ^2H", also space inserted around minus sign to read as "$\delta^2H - 8\,\delta^{18}O$

Page 96, lines 49–50 (right column), bad line break adjusted to keep "δ" and "^{2}H" together to read as "δ^2H"

Page 97, lines 7 and 8 (left column), bad line break adjusted to keep comma and NO_3^- together

Page 98, lines 18 and 19 (right column): bad line break adjusted to keep "NO_3" and "$^-$" together to read as "NO_3^-"

Page 98, lines 22 and 23 (right column): bad line break adjusted to keep "L" and "$^{-1}$" together to read as "L^{-1}"

MIX
Papier aus verantwortungsvollen Quellen
Paper from responsible sources
FSC® C105338

If you have any concerns about our products,
you can contact us on
ProductSafety@springernature.com

In case Publisher is established outside the EU,
the EU authorized representative is:
Springer Nature Customer Service Center GmbH
Europaplatz 3, 69115 Heidelberg, Germany

Printed by Libri Plureos GmbH
in Hamburg, Germany